T0292158

CAMBRIDGE LIBRARY COLLECTION

Books of enduring scholarly value

Darwin

Two hundred years after his birth and 150 years after the publication of 'On the Origin of Species', Charles Darwin and his theories are still the focus of worldwide attention. This series offers not only works by Darwin, but also the writings of his mentors in Cambridge and elsewhere, and a survey of the impassioned scientific, philosophical and theological debates sparked by his 'dangerous idea'.

On the Genesis of Species

'A distinguished zoologist, Mr St George Mivart, has recently collected all the objections which have ever been advanced by myself and others against the theory of natural selection, as propounded by Mr Wallace and myself, and has illustrated them with admirable art and force. When thus marshalled, they make a formidable array.' - Charles Darwin, in The Origin of Species by Means of Natural Selection (in its 6th edn, p. 176, published in 1872). St George Jackson Mivart was an eminent biologist, who was at first an advocate for natural selection and later a passionate opponent. In this beautifully illustrated 1871 text, Mivart raised objections to natural selection as a means for evolution. These included problems in explaining: 'incipient stages' of complex structures (e.g. the mammalian eye); the existence of similar structures of divergent origin; dramatic and rapid changes in form; the absence of transitional forms from the fossil record; and issues in geological distribution. Citing the giraffe's neck, the rattle of the snake, the whale's baleen and more, Mivart challenged Darwin's theory and argued for the necessity of an innate power - an intelligence - underlying all organic life. Mivart's book did not seriously undermine the concept of natural selection - Darwin and Huxley soon countered his 'formidable array' of arguments - but it helped move the debate forward. Sadly, it eventually led to a rift between Mivart and Darwin.

Cambridge University Press has long been a pioneer in the reissuing of out-of-print titles from its own backlist, producing digital reprints of books that are still sought after by scholars and students but could not be reprinted economically using traditional technology. The Cambridge Library Collection extends this activity to a wider range of books which are still of importance to researchers and professionals, either for the source material they contain, or as landmarks in the history of their academic discipline.

Drawing from the world-renowned collections in the Cambridge University Library, and guided by the advice of experts in each subject area, Cambridge University Press is using state-of-the-art scanning machines in its own Printing House to capture the content of each book selected for inclusion. The files are processed to give a consistently clear, crisp image, and the books finished to the high quality standard for which the Press is recognised around the world. The latest print-on-demand technology ensures that the books will remain available indefinitely, and that orders for single or multiple copies can quickly be supplied.

The Cambridge Library Collection will bring back to life books of enduring scholarly value (including out-of-copyright works originally issued by other publishers) across a wide range of disciplines in the humanities and social sciences and in science and technology.

On the Genesis of Species

ST GEORGE JACKSON MIVART

CAMBRIDGE
UNIVERSITY PRESS

CAMBRIDGE UNIVERSITY PRESS

Cambridge New York Melbourne Madrid Cape Town Singapore São Paolo Delhi

Published in the United States of America by Cambridge University Press, New York

www.cambridge.org
Information on this title: www.cambridge.org/9781108005524

© in this compilation Cambridge University Press 2009

This edition first published 1871
This digitally printed version 2009

ISBN 978-1-108-00552-4

ON THE GENESIS OF SPECIES.

ON THE

GENESIS OF SPECIES.

BY

ST. GEORGE MIVART, F.R.S.

London:

MACMILLAN AND CO.

1871.

TO

SIR HENRY HOLLAND, Bart., M.D.,

F.R.S., D.C.L., etc. etc.

My dear Sir Henry,

In giving myself the pleasure to dedicate, as I now do, this work to you, it is not my intention to identify you with any views of my own advocated in it. I simply avail myself of an opportunity of paying a tribute of esteem and regard to my earliest scientific friend—the first to encourage me in pursuing the study of nature.

I remain,

My dear Sir Henry,

Ever faithfully yours,

ST. GEORGE MIVART.

7, North Bank, Regent's Park,
December 8, 1870.

CONTENTS.

CHAPTER I

INTRODUCTORY

The problem of the genesis of species stated.—Nature of its probable solution.—Importance of the question.—Position here defended. —Statement of the DARWINIAN THEORY.—Its applicability to details of geographical distribution ; to rudimentary structures ; to homology ; to mimicry, &c.—Consequent utility of the theory.—Its wide acceptance.—Reasons for this other than, and in addition to, its scientific value. Its simplicity.—Its bearing on religious questions.—*Odium theologicum* and *odium antitheologicum.*—The antagonism supposed by many to exist between it and theology neither necessary nor universal. —Christian authorities in favour of evolution.—Mr. Darwin's "Animals and Plants under Domestication."—Difficulties of the Darwinian theory enumerated . . . . *Page* 1

CHAPTER II.

THE INCOMPETENCY OF "NATURAL SELECTION" TO ACCOUNT FOR THE INCIPIENT STAGES OF USEFUL STRUCTURES.

Mr. Darwin supposes that Natural Selection acts by slight variations.— These must be useful at once.—Difficulties as to the giraffe ; as to mimicry ; as to the heads of flat-fishes ; as to the origin and constancy of the vertebrate limbs ; as to whalebone ; as to the young kangaroo ;

as to sea-urchins ; as to certain processes of metamorphosis; as to the mammary gland ; as to certain ape characters ; as to the rattle-snake and cobra ; as to the process of formation of the eye and ear ; as to the fully developed condition of the eye and ear ; as to the voice ; as to shell-fish ; as to orchids ; as to ants.—The necessity for the simultaneous modification of many individuals.—Summary and conclusion . . . . *Page* 23

CHAPTER III.

THE CO-EXISTENCE OF CLOSELY SIMILAR STRUCTURES OF DIVERSE ORIGIN.

Chances against concordant variations.—Examples of discordant ones.— Concordant variations not unlikely on a non-Darwinian evolutionary hypothesis.—Placental and implacental mammals.—Birds and reptiles.—Independent origins of similar sense organs.—The ear.—The eye.— Other coincidences.—Causes besides Natural Selection produce concordant variations in certain geographical regions.—Causes besides Natural Selection produce concordant variations in certain zoological and botanical groups.—There are homologous parts not genetically related.—Harmony in respect of the organic and inorganic worlds.—Summary and conclusion . . . . *Page* 63

CHAPTER IV.

MINUTE AND GRADUAL MODIFICATIONS.

There are difficulties as to minute modifications, even if not fortuitous.— Examples of sudden and considerable modifications of different kinds.— Professor Owen's view.—Mr. Wallace.—Professor Huxley.—Objections to sudden changes.—Labyrinthodont.—Potto.—Cetacea.—As to origin of bird's wing.—Tendrils of climbing plants.—Animals once supposed to be connecting links.—Early specialization of structure.—Macrauchenia.—Glyptodon.—Sabre-toothed tiger.—Conclusion *Page* 97

CHAPTER V.

AS TO SPECIFIC STABILITY.

What is meant by the phrase " specific stability ;" such stability to be
expected *a priori*, or else considerable changes at once.—Rapidly
increasing difficulty of intensifying race characters ; alleged causes of
this phenomenon ; probably an internal cause co-operates.—A certain
definiteness in variations.—Mr. Darwin admits the principle of specific
stability in certain cases of unequal variability.—The goose.—The
peacock.—The guinea fowl.—Exceptional causes of variation under
domestication.—Alleged tendency to reversion.—Instances.—Sterility
of hybrids.—Prepotency of pollen of same species, but of different
race.—Mortality in young gallinaceous hybrids.—A bar to intermixture
exists somewhere.—Guinea-pigs.—Summary and conclusion . *Page* 113

CHAPTER VI.

SPECIES AND TIME.

Two relations of species to time.—No evidence of past existence of minutely
intermediate forms when such might be expected *a priori*.—Bats,
Pterodactyles, Dinosauria, and Birds.—Ichthyosauria, Chelonia, and
Anoura.—Horse ancestry.—Labyrinthodonts and Trilobites.—Two sub-
divisions of the second relation of species to time.—Sir William Thomson's views.—Probable period required for ultimate specific evolution
from primitive ancestral forms.—Geometrical increase of time required
for rapidly multiplying increase of structural differences.—Proboscis
monkey.—Time required for deposition of strata necessary for Darwinian evolution. — High organization of Silurian forms of life.—
Absence of fossils in oldest rocks.—Summary and conclusion.
Page 128

CHAPTER VII.

SPECIES AND SPACE.

The geographical distribution of animals presents difficulties.—These not
insurmountable in themselves ; harmonize with other difficulties.—
Fresh-water fishes.—Forms common to Africa and India ; to Africa and

South America ; to China and Australia ; to North America and China ; to New Zealand and South America ; to South America and Tasmania ; to South America and Australia.—Pleurodont lizards.—Insectivorous mammals.—Similarity of European and South American frogs.—Analogy between European salmon and fishes of New Zealand, &c.—An ancient Antarctic continent probable.—Other modes of accounting for facts of distribution.—Independent origin of closely similar forms.—Conclusion . . . . *Page* 144

CHAPTER VIII.

HOMOLOGIES.

Animals made up of parts mutually related in various ways.—What homology is.—Its various kinds.—Serial homology.—Lateral homology.—Vertical homology.—Mr. Herbert Spencer's explanations.—An internal power necessary, as shown by facts of comparative anatomy.—Of teratology.—M. St. Hilaire.—Professor Burt Wilder.—Foot-wings.—Facts of pathology.—Mr. James Paget.—Dr. William Budd.—The existence of such an internal power of individual development diminishes the improbability of an analogous law of specific origination . *Page* 155

CHAPTER IX.

EVOLUTION AND ETHICS.

The origin of morals an inquiry not foreign to the subject of this book.—Modern utilitarian view as to that origin.—Mr. Darwin's speculation as to the origin of the abhorrence of incest.—Cause assigned by him insufficient.—Care of the aged and infirm opposed by "Natural Selection ;" also self-abnegation and asceticism.—Distinctness of the ideas right and useful.—Mr. John Stuart Mill.—Insufficiency of "Natural Selection " to account for the origin of the distinction between duty and profit.—Distinction of moral acts into material and formal.—No ground for believing that formal morality exists in brutes.—Evidence that it does exist in savages.—Facility with which savages may be misunderstood.—Objections as to diversity of customs.—Mr. Hutton's review of Mr. Herbert Spencer.—Anticipatory character of morals.— Sir John Lubbock's explanation.—Summary and conclusion . *Page* 188

CHAPTER X.

P .1 N G E N E S I S.

A provisional hypothesis supplementing "Natural Selection."—Statement of the hypothesis.—Difficulty as to multitude of gemmules.—As to certain modes of reproduction.—As to formations without the requisite gemmules.—Mr. Lewes and Professor Delpino.—Difficulty as to developmental force of gemmules.—As to their spontaneous fission.— Pangenesis and Vitalism.—Paradoxical reality.—Pangenesis scarcely superior to anterior hypotheses.—Buffon.—Owen.—Herbert Spencer.— Gemmules as mysterious as " physiological units."—Conclusion.
Page 208

CHAPTER XI.

S P E C I F I C G E N E S I S.

Review of the statements and arguments of preceding chapters.—Cumulative argument against predominant action of " Natural Selection."— Whether anything positive as well as negative can be enunciated.— Constancy of laws of nature does not necessarily imply constancy of specific evolution.—Possible exceptional stability of existing epoch.— Probability that an internal cause of change exists.—Innate powers somewhere must be accepted.—Symbolism of molecular action under vibrating impulses. Professor Owen's statement.—Statement of the Author's view.—It avoids the difficulties which oppose "Natural Selection."—It harmonizes apparently conflicting conceptions.—Summary and conclusion .   .   .   .   *Page* 220

CHAPTER XII.

THEOLOGY AND EVOLUTION.

Prejudiced opinions on the subject.—"Creation" sometimes denied from prejudice.—The unknowable.—Mr. Herbert Spencer's objections to theism ; to creation.—Meanings of term "creation."—Confusion from not distinguishing between "primary" and "derivative" creation.— Mr. Darwin's objections.--Bearing of Christianity on evolution.—Supposed opposition, the result of a misconception.—Theological authority not opposed to evolution.—St. Augustin.—St. Thomas Aquinas.— Certain consequences of want of flexibility of mind.—Reason and imagination.—The first cause and demonstration.—Parallel between Christianity and natural theology.—What evolution of species is.— Professor Agassiz.—Innate powers must be recognized.—Bearing of evolution on religious belief.—Professor Huxley.—Professor Owen.— Mr. Wallace.—Mr. Darwin.—*A priori* conception of Divine action.— Origin of man.—Absolute creation and dogma.—Mr. Wallace's view.— A supernatural origin for man's body not necessary.—Two orders of being in man.—Two modes of origin.—Harmony of the physical, hyperphysical, and supernatural.—Reconciliation of science and religion as regards evolution.—Conclusion . . . . . *Page* 243

INDEX . . . . . *Page* 289

LIST OF ILLUSTRATIONS.

PAGE

Leaf Butterfly in flight and repose (*from Mr. A. Wallace's "Malay Archipelago"*) 31

Walking-Leaf Insect 35

Pleuronectidæ, with the peculiarly placed eye in different positions (*from Dr. Traquair's paper in Linn. Soc. Trans.*, 1865) 37, 166

Mouth of Whale (*from Professor Owen's "Odontography"*) ... 40

Four plates of Baleen seen obliquely from within (*from Professor Owen's "Odontography"*) 41

Dugong 41, 175

Echinus or Sea Urchin 43, 167

Pedicellariæ of Echinus very much enlarged... 44

Rattlesnake 49

Cobra (*from Sir Andrew Smith's "Southern Africa"*) 50

Wingbones of Pterodactyle, Bat, and Bird (*from Mr. Andrew Murray's "Geographical Distribution of Mammals"*) ... 64, 130, 157

Skeleton of Flying-Dragon 65, 158

Centipede (*from a specimen in the Museum of the Royal College of Surgeons*) 66, 159

Teeth of Urotrichus and Perameles 68

The Archeopteryx (*from Professor Owen's "Anatomy of Vertebrata"*) ... ... 73, 132 xiv

LIST OF ILLUSTRATIONS.

PAGE

Cuttle-Fish 75, 141

Skeleton of Ichthyosaurus 78, 107, 132, 177

Cytheridea Torosa (*from Messrs. Brady and Robertson's paper in Ann. and Mag. of Nat. Hist.,* 1870) 79

A Polyzoon, with Bird's-head processes 80

Bird's-head processes greatly enlarged 81

Antechinus Minutissimus and Mus Delicatulus (*from Mr. Andrew Murray's " Geographical Distribution of Mammals"*) 82

Outlines of Wings of Butterflies of Celebes compared with those of allied species elsewhere 86

Great Shielded Grasshopper 89

The Six-shafted Bird of Paradise 90

The Long-tailed Bird of Paradise 91

The Red Bird of Paradise 92

Horned Flies 93

The Magnificent Bird of Paradise 93

(*The above seven figures are from Mr. A. Wallace's " Malay Archipelago."*)

Much enlarged horizontal Section of the Tooth of a Labyrinthodon (*from Professor Owen's " Odontography"*) 104

Hand of the Potto (*from life*) 105

Skeleton of Plesiosaurus 106, 133

The Aye-Aye (*from Trans. of Zool. Soc.*) 108

Dentition of Sabre-toothed Tiger (*from Professor Owen's " Odontography"*) 110

Trilobite 135, 171

Inner side of Lower Jaw of Pleurodont Lizard (*from Professor Owen's " Odontography"*) 148

Solenodon (*from Berlin Trans.*) 149

Tarsal Bones of Galago and Cheirogaleus (*from Proc. Zool. Soc.*) ... 159

Squilla 160

Parts of the Skeleton of the Lobster 161

PAGE

Spine of Galago Allenii (*from Proc. Zool. Soc.*)   162

Vertebræ of Axolotl (*from Proc. Zool. Soc.*)...   165

Annelid undergoing spontaneous fission 169, 211

Aard-Vark (*Orycteropus capensis*)   174

Pangolin (*Manis*)   175

Skeleton of Manus and Pes of a Tailed Batrachian (*from Professor Gegenbaur's " Tarsus and Carpus"*)   178

Flexor Muscles of Hand of Nycticetus (*from Proc. Zool. Soc.*)   ...   180

The Fibres of Corti   279

THE GENESIS OF SPECIES.

CHAPTER I.

INTRODUCTORY.

The problem of the genesis of species stated.—Nature of its probable
solution.—Importance of the question.—Position here defended.—
Statement of the DARWINIAN THEORY.—Its applicability to details
of geographical distribution ; to rudimentary structures ; to homology ; to mimicry, &c.—Consequent utility of the theory.—Its wide
acceptance.—Reasons for this, other than, and in addition to, its scientific value.—Its simplicity.—Its bearing on religious questions.—*Odium
theologicum* and *odium antitheologicum.*—The antagonism supposed by
many to exist between it and theology neither necessary nor universal.
—Christian authorities in favour of evolution.—Mr. Darwin's "Animals
and Plants under Domestication."—Difficulties of the Darwinian
theory enumerated.

THE great problem which has so long exercised the minds of
naturalists, namely, that concerning the origin of different kinds
of animals and plants, seems at last to be fairly on the road
to receive—perhaps at no very distant future—as satisfactory a
solution as it can well have.

But the problem presents peculiar difficulties. The birth of
a "species" has often been compared with that of an "individual." The origin, however, of even an individual animal or
plant (that which determines an embryo to evolve itself,—as,

e.g., a spider rather than a beetle, a rose-plant rather than a pear) is shrouded in obscurity. *A fortiori* must this be the case with the origin of a "species."

Moreover, the analogy between a "species" and an "individual" is a very incomplete one. The word "individual" denotes a concrete whole with a real, separate, and distinct existence. The word "species," on the other hand, denotes a peculiar congeries of characters, innate powers and qualities, and a certain nature realized indeed in individuals, but having no separate existence, except ideally as a thought in some mind.

Thus the birth of a "species" can only be compared metaphorically, and very imperfectly, with that of an "individual."

Individuals *as individuals*, actually and directly produce and bring forth other individuals ; but no "congeries of characters" no "common nature" *as such*, can directly bring forth another "common nature," because, *per se*, it has no existence (other than ideal) apart from the individuals in which it is manifested.

The problem then is, "by what combination of natural laws does a new 'common nature' appear upon the scene of realized existence?" *i.e.* how is an individual embodying such new characters produced ?

For the approximation we have of late made towards the solution of this problem, we are mainly indebted to the invaluable labours and active brains of Charles Darwin and Alfred Wallace.

Nevertheless, important as has been the impulse and direction given by those writers to both our observations and speculations, the solution will not (if the views here advocated are correct) ultimately present that aspect and character with which it has issued from the hands of those writers.

Neither, most certainly, will that solution agree in appearance or substance with the more or less crude conceptions which have been put forth by most of the opponents of Messrs. Darwin and Wallace.

Rather, judging from the more recent manifestations of thought on opposite sides, we may expect the development of some *tertium quid*—the resultant of forces coming from different quarters, and not coinciding in direction with any one of them.

As error is almost always partial truth, and so consists in the exaggeration or distortion of one verity by the suppression of another which qualifies and modifies the former, we may hope, by the synthesis of the truths contended for by various advocates, to arrive at the one conciliating reality.

Signs of this conciliation are not wanting : opposite scientific views, opposite philosophical conceptions, and opposite religious beliefs, are rapidly tending by their vigorous conflict to evolve such a systematic and comprehensive view of the genesis of species as will completely harmonize with the teachings of science, philosophy, and religion.

To endeavour to add one stone to this temple of concord, to try and remove a few of the misconceptions and mutual misunderstandings which oppose harmonious action, is the aim and endeavour of the present work. This aim it is hoped to attain, not by shirking difficulties, but analysing them, and by endeavouring to dig down to the common root which supports and unites diverging stems of truth.

It cannot but be a gain when the labourers in the three fields above mentioned, namely, science, philosophy, and religion, shall fully recognize this harmony. Then the energy too often spent in futile controversy, or withheld through prejudice, may be profitably and reciprocally exercised for the mutual benefit of all.

Remarkable is the rapidity with which an interest in the question of specific origination has spread. But a few years ago it scarcely occupied the minds of any but naturalists. Then the crude theory put forth by Lamarck, and by his English interpreter the author of the "Vestiges of Creation," had rather discredited than helped on a belief in organic evolution—a belief,

that is, in new kinds being produced from older ones by the
ordinary and constant operation of natural laws. Now, however, this belief is widely diffused. Indeed, there are few
drawing-rooms where it is not the subject of occasional discussion, and artisans and schoolboys have their views as to the
permanence of organic forms. Moreover, the reception of this
doctrine tends actually, though by no means necessarily, to be
accompanied by certain beliefs with regard to quite distinct
and very momentous subject-matter. So that the question of
the "Genesis of Species" is not only one of great interest,
but also of much importance.

But though the calm and thorough consideration of this
matter is at the present moment exceedingly desirable, yet the
actual importance of the question itself as to its consequences
in the domain of theology has been strangely exaggerated by
many, both of its opponents and supporters. This is especially
the case with that form of the evolution theory which is associated with the name of Mr. Darwin; and yet neither the refutation nor the demonstration of that doctrine would be necessarily accompanied by the results which are hoped for by one
party and dreaded by another.

The general theory of evolution has indeed for some time
past steadily gained ground, and it may be safely predicted
that the number of facts which can be brought forward in its
support will, in a few years, be vastly augmented. But the
prevalence of this theory need alarm no one, for it is, without any doubt, perfectly consistent with strictest and most
orthodox Christian theology. Moreover, it is not altogether
without obscurities, and cannot yet be considered as fully
demonstrated.

The special Darwinian hypothesis, however, is beset with
certain scientific difficulties, which must by no means be ignored, and some of which, I venture to think, are absolutely
insuperable. What Darwinism or "Natural Selection" is, will

be shortly explained ; but before doing so, I think it well to state the object of this book, and the view taken up and defended in it. It is its object to maintain the position that "Natural Selection" acts, and indeed must act, but that still, in order that we may be able to account for the production of known kinds of animals and plants, it requires to be supplemented by the action of some other natural law or laws as yet undiscovered.[1] Also, that the consequences which have been drawn from Evolution, whether exclusively Darwinian or not, to the prejudice of religion, by no means follow from it, and are in fact illegitimate.

The Darwinian theory of "Natural Selection" may be shortly stated thus :[2]—

Every kind of animal and plant tends to increase in numbers in a geometrical progression.

Every kind of animal and plant transmits a general likeness, with individual differences, to its offspring.

Every individual may present minute variations of any kind and in any direction.

Past time has been practically infinite.

Every individual has to endure a very severe struggle for existence, owing to the tendency to geometrical increase of all kinds of animals and plants, while the total animal and vegetable population (man and his agency excepted) remains almost stationary.

Thus, every variation of a kind tending to save the life of the individual possessing it, or to enable it more surely to propagate its kind, will in the long run be preserved, and will transmit its favourable peculiarity to some of its offspring,

[1] In the last edition of the "Origin of Species" (1869) Mr. Darwin himself admits that "Natural Selection" has not been the exclusive means of modification, though he still contends it has been the most important one.

[2] See Mr. Wallace's recent work, entitled "Contributions to the Theory of Natural Selection," where, at p. 302, it is very well and shortly stated.

which peculiarity will thus become intensified till it reaches the maximum degree of utility. On the other hand, individuals presenting unfavourable peculiarities will be ruthlessly destroyed. The action of this law of Natural Selection may thus be well represented by the convenient expression "survival of the fittest."[1]

Now this conception of Mr. Darwin's is perhaps the most interesting theory, in relation to natural science, which has been promulgated during the present century. Remarkable, indeed, is the way in which it groups together such a vast and varied series of biological[2] facts, and even paradoxes, which it appears more or less clearly to explain, as the following instances will show. By this theory of "Natural Selection," light is thrown on the more singular facts relating to the geographical distribution of animals and plants; for example, on the resemblance between the past and present inhabitants of different parts of the earth's surface. Thus in Australia remains have been found of creatures closely allied to kangaroos and other kinds of pouched beasts, which in the present day exist nowhere but in the Australian region. Similarly in South America, and nowhere else, are found sloths and armadillos, and in that same part of the world have been discovered bones of animals different indeed from existing sloths and armadillos, but yet much more nearly related to them than to any other kinds whatever. Such coincidences between the existing and antecedent geographical distribution of forms are numerous. Again, "Natural Selection" serves to explain the circumstance that often in adjacent islands we find animals closely resembling, and appearing to represent, each other; while if certain of these islands show signs (by depth of surrounding sea or what not) of more ancient separation, the

[1] "Natural Selection" is happily so termed by Mr. Herbert Spencer in his "Principles of Biology."

[2] Biology is the science of life. It contains zoology, or the science of animals, and botany, or that of plants.

animals inhabiting them exhibit a corresponding divergence.[1] The explanation consists in representing the forms inhabiting the islands as being the modified descendants of a common stock, the modification being greatest where the separation has been the most prolonged.

"Rudimentary structures" also receive an explanation by means of this theory. These structures are parts which are apparently functionless and useless where they occur, but which represent similar parts of large size and functional importance in other animals. Examples of such "rudimentary structures" are the foetal teeth of whales, and of the front part of the jaw of ruminating quadrupeds. These foetal structures are minute in size, and never cut the gum, but are reabsorbed without ever coming into use, while no other teeth succeed them or represent them in the adult condition of those animals. The mammary glands of all male beasts constitute another example, as also does the wing of the apteryx—a New Zealand bird utterly incapable of flight, and with the wing in a quite rudimentary condition (whence the name of the animal). Yet this rudimentary wing contains bones which are miniature representatives of the ordinary wing-bones of birds of flight. Now, the presence of these useless bones and teeth is explained if they may be considered as actually being the inherited diminished representatives of parts of large size and functional importance in the remote ancestors of these various animals.

Again, the singular facts of "homology" are capable of a similar explanation. "Homology," is the name applied to the investigation of those profound resemblances which have so often been found to underlie superficial differences between animals of very different form and habit. Thus man, the horse, the whale, and the bat, all have the pectoral limb, whether it be the arm, or fore-leg, or paddle, or wing, formed on essentially the same type, though

---

[1] For very interesting examples, see Mr. Wallace's "Malay Archipelago."

the number and proportion of parts may more or less differ. Again, the butterfly and the shrimp, different as they are in appearance and mode of life, are yet constructed on the same common plan, of which they constitute diverging manifestations. No *a priori* reason is conceivable why such similarities should be necessary, but they are readily explicable on the assumption of a genetic relationship and affinity between the animals in question, assuming, that is, that they are the modified descendants of some ancient form—their common ancestor.

That remarkable series of changes which animals undergo before they attain their adult condition, which is called their process of development, and during which they more or less closely resemble other animals during the early stages of the same process, has also great light thrown on it from the same source. The question as to the singularly complex resemblances borne by every adult animal and plant to a certain number of other animals and plants—resemblances by means of which the adopted zoological and botanical systems of classification have been possible—finds its solution in a similar manner, classification becoming the expression of a genealogical relationship. Finally, by this theory—and as yet by this alone—can any explanation be given of that extraordinary phenomenon which is metaphorically termed *mimicry*. Mimicry is a close and striking, yet superficial resemblance borne by some animal or plant to some other, perhaps very different, animal or plant. The "walking leaf" (an insect belonging to the grasshopper and cricket order) is a well-known and conspicuous instance of the assumption by an animal of the appearance of a vegetable structure (see illustration on p. 35); and the bee, fly, and spider orchids are familiar examples of a converse resemblance. Birds, butterflies, reptiles, and even fish, seem to bear in certain instances a similarly striking resemblance to other birds, butterflies, reptiles, and fish, of altogether distinct kinds. The explanation of this matter which " Natural Selection" offers to animals, is that certain varieties of one kind have found exemption from persecution in consequence of an accidental resemblance which such varieties have exhibited to animals of another kind, or to plants ; and that they were thus preserved, and the degree of resemblance was continually augmented in their descendants.   As to plants, the explanation offered by this theory might perhaps be that varieties of plants which presented a certain superficial resemblance in their flowers to insects, have thereby been helped to propagate their kind, the visit of certain insects being useful or indispensable to the fertilization of many flowers.

We have thus a whole series of important facts which "Natural Selection" helps us to understand and co-ordinate. And not only are all these diverse facts strung together, as it were, by the theory in question; not only does it explain the development of the complex instincts of the beaver, the cuckoo, the bee, and the ant, as also the dazzling brilliancy of the humming-bird, the glowing tail and neck of the peacock, and the melody of the nightingale ; the perfume of the rose and the violet, the brilliancy of the tulip and the sweetness of the nectar of flowers ; not only does it help us to understand all these, but serves as a basis of future research and of inference from the known to the unknown, and it guides the investigator to the discovery of new facts which, when ascertained, it seems also able to co-ordinate.[1]   Nay, "Natural Selection" seems capable of application not only to the building up of the smallest and most insignificant organisms, but even of extension beyond the biological domain altogether, so as possibly to have relation to

[1] See Müller's work, "Für Darwin," lately translated into English by Mr. Dallas. Mr. Wallace also predicts the discovery, in Madagascar, of a hawk-moth with an enormously long proboscis, and he does this on account of the discovery there of an orchid with a nectary from ten to fourteen inches in length.   See *Quarterly Journal of Science*, October 1867, and "Natural Selection," p. 275.

the stable equilibrium of the solar system itself, and even of the whole sidereal universe. Thus, whether this theory be true or false, all lovers of natural science should acknowledge a deep debt of gratitude to Messrs. Darwin and Wallace, on account of its practical utility. But the utility of a theory by no means implies its truth. What do we not owe, for example, to the labours of the Alchemists? The emission theory of light, again, has been pregnant with valuable results, as still is the Atomic theory, and others which will readily suggest themselves.

With regard to Mr. Darwin (with whose name, on account of the noble self-abnegation of Mr. Wallace, the theory is in general exclusively associated), his friends may heartily congratulate him on the fact that he is one of the few exceptions to the rule respecting the non-appreciation of a prophet in his own country. It would be difficult to name another living labourer in the field of physical science who has excited an interest so widespread, and given rise to so much praise, gathering round him, as he has done, a chorus of more or less completely acquiescing disciples, themselves masters in science, and each the representative of a crowd of enthusiastic followers.

Such is the Darwinian theory of "Natural Selection," such are the more remarkable facts which it is potent to explain, and such is the reception it has met with in the world. A few words now as to the reasons for the very widespread interest it has awakened, and the keenness with which the theory has been both advocated and combated.

The important bearing it has on such an extensive range of scientific facts, its utility, and the vast knowledge and great ingenuity of its promulgator, are enough to account for the heartiness of its reception by those learned in natural history. But quite other causes have concurred to produce the general and higher degree of interest felt in the theory beside the readiness with which it harmonizes with biological facts. These latter could only be appreciated by physiologists, zoologists, and botanists; whereas the Darwinian theory, so novel and so startling, has found a cloud of advocates and opponents beyond and outside the world of physical science.

In the first place, it was inevitable that a great crowd of half-educated men and shallow thinkers should accept with eagerness the theory of " Natural Selection," or rather what they think to be such (for few things are more remarkable than the way in which it has been misunderstood), on account of a certain characteristic it has in common with other theories; which should not be mentioned in the same breath with it, except, as now, with the accompaniment of protest and apology. We refer to its remarkable simplicity, and the ready way in which phenomena the most complex appear explicable by a cause for the comprehension of which laborious and persevering efforts are not required, but which may be represented by the simple phrase "survival of the fittest." With nothing more than this, can, on the Darwinian theory, all the most intricate facts of distribution and affinity, form, and colour, be accounted for; as well the most complex instincts and the most admirable adjustments, such as those of the human eye and ear. It is in great measure then, owing to this supposed simplicity, and to a belief in its being yet easier and more simple than it is, that Darwinism, however imperfectly understood, has become a subject for general conversation, and has been able thus widely to increase a certain knowledge of biological matters; and this excitation of interest in quarters where otherwise it would have been entirely wanting, is an additional motive for gratitude on the part of naturalists to the authors of the new theory. At the same time it must be admitted that a similar "simplicity" —the apparently easy explanation of complex phenomena— also constitutes the charm of such matters as hydropathy and phrenology, in the eyes of the unlearned or half-educated public. It is indeed *the* charm of all those seeming "short cuts" to knowledge, by which the labour of mastering scientific details is spared to those who yet believe that without such labour they can attain all the most valuable results of scientific research. It is not, of course, for a moment meant to imply that its "simplicity" tells at all against "Natural Selection," but only that the actual or supposed possession of that quality is a strong reason for the wide and somewhat hasty acceptance of the theory, whether it be true or not.

In the second place, it was inevitable that a theory appearing to have very grave relations with questions of the last importance and interest to man, that is, with questions of religious belief, should call up an army of assailants and defenders. Nor have the supporters of the theory much reason, in many cases, to blame the more or less unskilful and hasty attacks of adversaries, seeing that those attacks have been in great part due to the unskilful and perverse advocacy of the cause on the part of some of its adherents. If the *odium theologicum* has inspired some of its opponents, it is undeniable that the *odium antitheologicum* has possessed not a few of its supporters. It is true (and in appreciating some of Mr. Darwin's expressions it should never be forgotten) that the theory has been both at its first promulgation and since vehemently attacked and denounced as unchristian, nay, as necessarily atheistic; but it is not less true that it has been made use of as a weapon of offence by irreligious writers, and has been again and again, especially in continental Europe, thrown, as it were, in the face of believers, with sneers and contumely. When we recollect the warmth with which what he thought was Darwinism was advocated by such a writer as Professor Vogt, one cause of his zeal was not far to seek—a zeal, by the way, certainly not "according to knowledge;" for few conceptions could have been more conflicting with true Darwinism than the theory he formerly maintained, but has since abandoned, viz. that the men of the Old World were descended from African and Asiatic apes, while, similarly, the American apes were the progenitors of the human beings of the New World.

The cause of this palpable error in a too eager disciple one might hope was not anxiety to snatch up all or any arms available against Christianity, were it not for the tone unhappily adopted by this author. But it is unfortunately quite impossible to mistake his meaning and intention, for he is a writer whose offensiveness is gross, while it is sometimes almost surpassed by an amazing shallowness. Of course, as might fully be expected, he adopts and reproduces the absurdly trivial objections to absolute morality drawn from differences in national customs.[1] And he seems to have as little conception of the distinction between "formally" moral actions and those which are only "materially" moral, as of that between the *verbum mentale* and the *verbum oris*. As an example of his onesidedness, it may be remarked that he compares the skulls of the American monkeys (*Cebus apella* and *C. albifrons*) with the intention of showing that man is of several distinct species, because skulls of different men are less alike than are those of these two monkeys; and he does this regardless of how the skulls of domestic animals (with which it is far more legitimate to compare races of men than with wild kinds), *e.g.* of different dogs or pigeons, tell precisely in the opposite direction. Regardless also of the fact that perhaps no genus of monkeys is in a more unsatisfactory state as to the determination of its different kinds than the genus chosen by him for illustration. This is so much the case that J. A. Wagner (in his supplement to Schreber's great work on Beasts) at first included all the kinds in a single species.

As to the strength of his prejudice and his regretable coarseness, one quotation will be enough to display both. Speaking of certain early Christian missionaries, he says,[2] " It is not so very improbable that the new religion, before which the flourishing Roman civilization relapsed into a state of barbarism, should have

---

[1] "Lectures on Man," translated by the Antaropological Society, 1864, p. 229. [2] Ibid. p. 378.

been introduced by people in whose skulls the anatomist finds simious characters so well developed, and in which the phrenologist finds the organ of veneration so much enlarged. I shall, in the meanwhile, call these simious narrow skulls of Switzerland 'Apostle skulls,' as I imagine that in life they must have resembled the type of Peter, the Apostle, as represented in Byzantine-Nazarene art."

In face of such a spirit, can it be wondered at that disputants have grown warm? Moreover, in estimating the vehemence of the opposition which has been offered, it should be borne in mind that the views defended by religious writers are, or should be, all-important in their eyes. They could not be expected to view with equanimity the destruction in many minds of "theology, natural and revealed, psychology, and metaphysics;" nor to weigh with calm and frigid impartiality arguments which seemed to them to be fraught with results of the highest moment to mankind, and, therefore, imposing on their consciences strenuous opposition as a first duty. Cool judicial impartiality in them would have been a sign perhaps of intellectual gifts, but also of a more important deficiency of generous emotion.

It is easy to complain of the onesidedness of many of those who oppose Darwinism in the interest of orthodoxy; but not at all less patent is the intolerance and narrow-mindedness of some of those who advocate it, avowedly or covertly, in the interest of heterodoxy. This hastiness of rejection or acceptance, determined by ulterior consequences believed to attach to "Natural Selection," is unfortunately in part to be accounted for by some expressions and a certain tone to be found in Mr. Darwin's writings. That his expressions, however, are not always to be construed literally is manifest. His frequent use metaphorically of the expressions, "contrivance," for example, and "purpose," has elicited, from the Duke of Argyll and others, criticisms which fail to tell against their opponent, because such expressions are, in Mr. Darwin's writings, merely figurative—metaphors, and nothing more.

It may be hoped, then, that a similar looseness of expression will account for passages of a directly opposite tendency to that of his theistic metaphors.

Moreover, it must not be forgotten that he frequently uses that absolutely theological term, " the Creator," and that he has retained in all the editions of his " Origin of Species " an expression which has been much criticised. He speaks " of life, with its several powers, having been originally breathed by the Creator into a few forms, or into one."[1] This is merely mentioned in justice to Mr. Darwin, and by no means because it is a position which this book is intended to support. For, from Mr. Darwin's usual mode of speaking, it appears that by such divine action he means a supernatural intervention, whereas it is here contended that throughout the whole process of physical evolution—the first manifestation of life included —*supernatural* action is assuredly not to be looked for.

Again, in justice to Mr. Darwin, it may be observed that he is addressing the general public, and opposing the ordinary and common objections of popular religionists, who have inveighed against " Evolution " and " Natural Selection " as atheistic, impious, and directly conflicting with the dogma of creation.

Still, in so important a matter, it is to be regretted that he did not take the trouble to distinguish between such merely popular views and those which repose upon some more venerable authority. Mr. John Stuart Mill has replied to similar critics, and shown that the assertion that his philosophy is irreconcilable with theism is unfounded ; and it would have been better if Mr. Darwin had dealt in the same manner with some of his assailants, and shown the futility of certain of their

[1] See Fifth Edition, 1869, p. 579.

objections when viewed from a more elevated religious stand-
point.   Instead of so doing, he seems to adopt the narrowest
notions of his opponents, and, far from endeavouring to expand
them, appears to wish to endorse them, and to lend to them the
weight of his authority.   It is thus that Mr. Darwin seems to
admit and assume that the idea of "creation" necessitates a
belief in an interference with, or dispensation of, natural laws,
and that "creation" must be accompanied by arbitrary and
unorderly phenomena.   None but the crudest conceptions are
placed by him to the credit of supporters of the dogma of
creation, and it is constantly asserted that they, to be consistent,
must offer "creative fiats" as explanations of physical phenomena, and be guilty of numerous other such absurdities.   It
is impossible, therefore, to acquit Mr. Darwin of at least a
certain carelessness in this matter ; and the result is, he has the
appearance of opposing ideas which he gives no clear evidence
of having ever fully appreciated.   He is far from being alone
in this, and perhaps merely takes up and reiterates, without
much consideration, assertions previously assumed by others.
Nothing could be further from Mr. Darwin's mind than any,
however small, intentional misrepresentation ; and it is therefore
the more unfortunate that he should not have shown any appreciation of a position opposed to his own other than that gross
and crude one which he combats so superfluously—that he
should appear, even for a moment, to be one of those, of whom
there are far too many, who first misrepresent their adversary's
view, and then elaborately refute it ; who, in fact, erect a doll
utterly incapable of self-defence and then, with a flourish of
trumpets and many vigorous strokes, overthrow the helpless
dummy they had previously raised.

   This is what many do who more or less distinctly oppose
theism in the interests, as they believe, of physical science ;
and they often represent, amongst other things, a gross and
narrow anthropomorphism as the necessary consequence of views

opposed to those which they themselves advocate. Mr. Darwin and others may perhaps be excused if they have not devoted much time to the study of Christian philosophy; but they have no right to assume or accept, without careful examination, as an unquestioned fact, that in that philosophy there is a necessary antagonism between the two ideas, "creation" and "evolution," as applied to organic forms.

It is notorious and patent to all who choose to seek, that many distinguished Christian thinkers have accepted and do accept both ideas, *i. e.* both "creation" and "evolution."

As much as ten years ago, an eminently Christian writer observed : " The creationist theory does not necessitate the perpetual search after manifestations of miraculous powers and perpetual 'catastrophes.' Creation is not a miraculous interference with the laws of nature, but the very institution of those laws. Law and regularity, not arbitrary intervention, was the patristic ideal of creation. With this notion, they admitted without difficulty the most surprising origin of living creatures, provided it took place by *law.* They held that when God said, ' Let the waters produce,' ' Let the earth produce,' He conferred forces on the elements of earth and water, which enabled them naturally to produce the various species of organic beings. This power, they thought, remains attached to the elements throughout all time."[1] The same writer quotes St. Augustine and St. Thomas Aquinas, to the effect that, "in the institution of nature we do not look for miracles, but for the laws of nature."[2] And, again, St. Basil,[3] speaks of the continued operation of natural laws in the production of all organisms.

[1] *The Rambler,* March 1860, vol. xii. p. 372.
[2] "In primâ institutione naturæ non quæritur miraculum, sed quid natura rerum habeat, ut Augustinus dicit, lib. ii. sup. Gen. ad lit. c. 1." (St. Thomas, Sum. I^æ. lxvii. 4, ad 3.)
[3] "Hexaem." Hom. ix. p. 81.

So much for writers of early and mediæval times. As to the present day, the Author can confidently affirm that there are many as well versed in theology as Mr. Darwin is in his own department of natural knowledge, who would not be disturbed by the thorough demonstration of his theory. Nay, they would not even be in the least painfully affected at witnessing the generation of animals of complex organization by the skilful artificial arrangement of natural forces, and the production, in the future, of a fish, by means analogous to those by which we now produce urea.

And this because they know that the possibility of such phenomena, though by no means actually foreseen, has yet been fully provided for in the old philosophy centuries before Darwin, or even before Bacon, and that their place in the system can be at once assigned them without even disturbing its order or marring its harmony.

Moreover, the old tradition in this respect has never been abandoned, however much it may have been ignored or neglected by some modern writers. In proof of this it may be observed that perhaps no post-mediæval theologian has a wider reception amongst Christians throughout the world than Suarez, who has a separate section[1] in opposition to those who maintain the distinct creation of the various kinds—or substantial forms—of organic life.

But the consideration of this matter must be deferred for the present, and the question of evolution, whether Darwinian or other, be first gone into. It is proposed, after that has been done, to return to this subject (here merely alluded to), and to consider at some length the bearing of " Evolution," whether Darwinian or non-Darwinian, upon " Creation and Theism."

Now we will revert simply to the consideration of the theory of " Natural Selection " itself.

[1] Suarez, Metaphysica. Edition Vivés. Paris, 1868. Vol. I. Disputatio xv. § 2.

Whatever may have hitherto been the amount of acceptance that this theory has met with, all, I think, anticipated that the appearance of Mr. Darwin's large and careful work on " Animals and Plants under Domestication " could but further increase that acceptance. It is, however, somewhat problematical how far such anticipations will be realized. The newer book seems to add after all but little in support of the theory, and to leave most, if not all, its difficulties exactly where they were. It is a question, also, whether the hypothesis of " Pangenesis " [1] may not be found rather to encumber than to support the theory it was intended to subserve. However, the work in question treats only of domestic animals, and probably the next instalment will address itself more vigorously and directly to the difficulties which seem to us yet to bar the way to a complete acceptance of the doctrine.

If the theory of Natural Selection can be shown to be quite insufficient to explain any considerable number of important phenomena connected with the origin of species, that theory, as *the* explanation, must be considered as provisionally discredited.

If other causes than Natural (including sexual) Selection can be proved to have acted—if variation can in any cases be proved to be subject to certain determinations in special directions by other means than Natural Selection, it then becomes probable *a priori* that it is so in others, and that Natural Selection

---

[1] " Pangenesis " is the name of the new theory proposed by Mr. Darwin, in order to account for various obscure physiological facts, such, *e.g.*, as the occasional reproduction, by individuals, of parts which they have lost ; the appearance in offspring of parental, and sometimes of remote ancestral, characters, &c. It accounts for these phenomena by supposing that every creature possesses countless indefinitely-minute organic atoms, termed " gemmules," which atoms are supposed to be generated in every part of every organ, to be in constant circulation about the body, and to have the power of reproduction. Moreover, atoms from every part are supposed to be stored in the generative products.

depends upon, and only supplements, such means, which conception is opposed to the pure Darwinian position.

Now it is certain, *a priori*, that variation is obedient to some law and therefore that "Natural Selection" itself must be capable of being subsumed into some higher law; and it is evident, I believe, *a posteriori*, that Natural Selection is, at the very least, aided and supplemented by some other agency.

Admitting, then, organic and other evolution, and that new forms of animals and plants (new species, genera, &c.) have from time to time been evolved from preceding animals and plants, it follows, if the views here advocated are true, that this evolution has not taken place by the action of "Natural Selection" *alone*, but through it (amongst other influences) aided by the concurrent action of some other natural law or laws, at present undiscovered; and probably that the genesis of species takes place partly, perhaps mainly, through laws which may be most conveniently spoken of as special powers and tendencies existing in each organism; and partly through influences exerted on each by surrounding conditions and agencies organic and inorganic, terrestrial and cosmical, among which the "survival of the fittest" plays a certain but subordinate part.

The theory of "Natural Selection" may (though it need not) be taken in such a way as to lead men to regard the present organic world as formed, so to speak, *accidentally*, beautiful and wonderful as is confessedly the hap-hazard result. The same may perhaps be said with regard to the system advocated by Mr. Herbert Spencer, who, however, also relegates "Natural Selection" to a subordinate *rôle*. The view here advocated, on the other hand, regards the whole organic world as arising and going forward in one harmonious development similar to that which displays itself in the growth and action of each separate individual organism. It also regards each such separate organism as the expression of powers and tendencies not to be accounted for by " Natural Selection " alone, or even by that together with merely the direct influence of surrounding conditions.

The difficulties which appear to oppose themselves to the reception of " Natural Selection" or " the survival of the fittest," as the one explanation of the origin of species, have no doubt been already considered by Mr. Darwin. Nevertheless, it may be worth while to enumerate them, and to state the considerations which appear to give them weight; and there is no doubt but that a naturalist so candid and careful as the author of the theory in question, will feel obliged, rather than the reverse, by the suggestion of all the doubts and difficulties which can be brought against it.

What is to be brought forward may be summed up as follows :—

That "Natural Selection" is incompetent to account for the incipient stages of useful structures.

That it does not harmonize with the co-existence of closely similar structures of diverse origin.

That there are grounds for thinking that specific differences may be developed suddenly instead of gradually.

That the opinion that species have definite though very different limits to their variability is still tenable.

That certain fossil transitional forms are absent, which might have been expected to be present.

That some facts of geographical distribution supplement other difficulties.

That the objection drawn from the physiological difference between " species " and " races " still exists unrefuted.

That there are many remarkable phenomena in organic forms upon which " Natural Selection" throws no light whatever, but the explanations of which, if they could be attained, might throw light upon specific origination.

Besides these objections to the sufficiency of "Natural Selection," others may be brought against the hypothesis of "Pangenesis," which, professing as it does to explain great difficulties, seems to do so by presenting others not less great—almost to be the explanation of *obscurum per obscurius.*

CHAPTER II.

THE INCOMPETENCY OF "NATURAL SELECTION" TO ACCOUNT FOR THE INCIPIENT STAGES OF USEFUL STRUCTURES.

Mr. Darwin supposes that natural selection acts by slight variations.— These must be useful at once.—Difficulties as to the giraffe; as to mimicry; as to the heads of flat-fishes; as to the origin and constancy of the vertebrate limbs; as to whalebone; as to the young kangaroo; as to sea-urchins; as to certain processes of metamorphosis; as to the mammary gland; as to certain ape characters; as to the rattlesnake and cobra; as to the process of formation of the eye and ear; as to the fully developed condition of the eye and ear; as to the voice; as to shell-fish; as to orchids; as to ants.—The necessity for the simultaneous modification of many individuals.—Summary and conclusion.

"Natural Selection," simply and by itself, is potent to explain the maintenance or the further extension and development of favourable variations, which are at once sufficiently considerable to be useful from the first to the individual possessing them. But Natural Selection utterly fails to account for the conservation and development of the minute and rudimentary beginnings, the slight and infinitesimal commencements of structures, however useful those structures may afterwards become.

Now, it is distinctly enunciated by Mr. Darwin, that the spontaneous variations upon which his theory depends are individually slight, minute, and insensible. He says,[1] "Slight

---

[1] "Animals and Plants under Domestication," vol. ii. p. 192.

individual differences, however, suffice for the work, and are probably the sole differences which are effective in the production of new species." And again, after mentioning the frequent sudden appearances of domestic varieties, he speaks of " the false belief as to the similarity of natural species in this respect." [1] In his work on the "Origin of Species," he also observes, " Natural Selection acts only by the preservation and accumulation of small inherited modifications." [2] And " Natural Selection, if it be a true principle, will banish the belief . . . of any great and sudden modification in their structure." [3] Finally, he adds, " If it could be demonstrated that any complex organ existed, which could not possibly have been formed by numerous, successive, slight modifications, my theory would absolutely break down." [4]

Now the conservation of minute variations in many instances is, of course, plain and intelligible enough ; such, *e.g.*, as those which tend to promote the destructive faculties of beasts of prey on the one hand, or to facilitate the flight or concealment of the animals pursued on the other; provided always that these minute beginnings are of such a kind as really to have a certain efficiency, however small, in favour of the conservation of the individual possessing them ; and also provided that no unfavourable peculiarity in any other direction accompanies and neutralizes, in the struggle for life, the minute favourable variation.

But some of the cases which have been brought forward, and which have met with very general acceptance, seem less satisfactory when carefully analysed than they at first appear to be. Amongst these we may mention "the neck of the giraffe."

At first sight it would seem as though a better example in support of " Natural Selection" could hardly have been chosen.

---

[1] "Animals and Plants under Domestication," vol. ii. p. 414.
[2] "Origin of Species," 5th edit., 1869, p. 110.
[3] Ibid. p. 111.   [4] Ibid. p. 227.

Let the fact of the occurrence of occasional, severe droughts in the country which that animal has inhabited be granted. In that case, when the ground vegetation has been consumed, and the trees alone remain, it is plain that at such times only those individuals (of what we assume to be the nascent giraffe species) which were able to reach high up would be preserved, and would become the parents of the following generation, some individuals of which would, of course, inherit that high-reaching power which alone preserved their parents. Only the high-reaching issue of these high-reaching individuals would again, *cæteris paribus*, be preserved at the next drought, and would again transmit to their offspring their still loftier stature; and so on, from period to period, through æons of time, all the individuals tending to revert to the ancient shorter type of body, being ruthlessly destroyed at the occurrence of each drought.

(1.) But against this it may be said, in the first place, that the argument proves too much; for, on this supposition, many species must have tended to undergo a similar modification, and we ought to have at least several forms, similar to the giraffe, developed from different Ungulata.[1] A careful observer of animal life, who has long resided in South Africa, explored the interior, and lived in the giraffe country, has assured the Author that the giraffe has powers of locomotion and endurance fully equal to those possessed by any of the other Ungulata of that continent. It would seem, therefore, that some of these other Ungulates ought to have developed in a similar manner as to the neck, under pain of being starved, when the long neck of the giraffe was in its incipient stage.

To this criticism it has been objected that different kinds of animals are preserved, in the struggle for life, in very different

---

[1] The order *Ungulata* contains the hoofed beasts; that is, all oxen, deer, anteolpes, sheep, goats, camels, hogs, the hippopotamus, the different kinds of rhinoceros, the tapirs, horses, asses, zebras, quaggas, &c.

ways, and even that "high reaching" may be attained in more modes than one—as, for example, by the trunk of the elephant. This is, indeed, true, but then none of the African Ungulata [1] have, nor do they appear ever to have had, any proboscis whatsoever; nor have they acquired such a development as to allow them to rise on their hind limbs and graze on trees in a kangaroo-attitude, nor a power of climbing, nor, as far as known, any other modification tending to compensate for the comparative shortness of the neck.   Again, it may perhaps be said that leaf-eating forms are exceptional, and that therefore the struggle to attain high branches would not affect many Ungulates.   But surely, when these severe droughts necessary for the theory occur, the ground vegetation is supposed to be exhausted ; and, indeed, the giraffe is quite capable of feeding from off the ground. , So that, in these cases, the other Ungulata *must* have taken to leaf eating or have starved, and thus must have had any accidental long-necked varieties favoured and preserved exactly as the long-necked varieties of the giraffe are supposed to have been favoured and preserved.

The argument as to the different modes of preservation has been very well put by Mr. Wallace,[2] in reply to the objection that "colour, being dangerous, should not exist in nature." This objection appears similar to mine ; as I say that a giraffe neck, being needful, there should be many animals with it, while the objector noticed by Mr. Wallace says, "a dull colour being needful, all animals should be so coloured."   And Mr. Wallace shows in reply how porcupines, tortoises and mussels, very hard-coated bombadier beetles, stinging insects and nauseous-tasted caterpillars, can afford to be brilliant by the various means of active defence or passive protection

---

[1] The elephants of Africa and India, with their extinct allies, constitute the order *Proboscidea*, and do not belong to the Ungulata.

[2] See "Natural Selection," pp. 60—75.

they possess, other than obscure colouration. He says "the attitudes of some insects may also protect them, as the habit of turning up the tail by the harmless rove-beetles (Staphylinidæ) no doubt leads other animals, besides children, to the belief that they can sting. The curious attitude assumed by sphinx caterpillars is probably a safeguard, as well as the blood-red tentacles which can suddenly be thrown out from the neck by the caterpillars of all the true swallow-tailed butterflies."

But, because many different kinds of animals can elude the observation or defy the attack of enemies in a great variety of ways, it by no means follows that there are any similar number and variety of ways for attaining vegetable food in a country where all such food, other than the lofty branches of trees, has been for a time destroyed. In such a country we have a number of vegetable-feeding Ungulates, all of which present minute variations as to the length of the neck. If, as Mr. Darwin contends, the natural selection of these favourable variations has alone lengthened the neck of the giraffe by preserving it during droughts; similar variations, in similarly-feeding forms, at the same times, ought similarly to have been preserved and so lengthened the neck of some other Ungulates by similarly preserving them during the same droughts.

(2.) It may be also objected, that the power of reaching upwards, acquired by the lengthening of the neck and legs, must have necessitated a considerable increase in the entire size and mass of the body (larger bones requiring stronger and more voluminous muscles and tendons, and these again necessitating larger nerves, more capacious blood-vessels, &c.), and it is very problematical whether the disadvantages thence arising would not, in times of scarcity, more than counterbalance the advantages.

For a considerable increase in the supply of food would be requisite on account of this increase in size and mass, while at the same time there would be a certain decrease in strength;

for, as Mr. Herbert Spencer says, [1] " It is demonstrable that the excess of absorbed over expended nutriment must, other things equal, become less as the size of an animal becomes greater. In similarly-shaped bodies, the masses vary as the cubes of the dimensions ; whereas the strengths vary as the squares of the dimensions." . . . . " Supposing a creature which a year ago was one foot high, has now become two feet high, while it is unchanged in proportions and structure—what are the necessary concomitant changes that have taken place in it? It is eight times as heavy ; that is to say, it has to resist eight times the strain which gravitation puts on its structure; and in producing, as well as in arresting, every one of its movements, it has to overcome eight times the inertia. Meanwhile, the muscles and bones have severally increased their contractile and resisting powers, in proportion to the areas of their trans-verse sections ; and hence are severally but four times as strong as they were. Thus, while the creature has doubled in height, and while its ability to overcome forces has quadrupled, the forces it has to overcome have grown eight times as great. Hence, to raise its body through a given space, its muscles have to be contracted with twice the intensity, at a double cost of matter expended." Again, as to the cost at which nutriment is distributed through the body, and effete matters removed from it, " Each increment of growth being added at the periphery of an organism, the force expended in the transfer of matter must increase in a rapid progression—a progression more rapid than that of the mass."

There is yet another point. Vast as may have been the time during which the process of evolution has continued, it is never-theless not infinite. Yet, as every kind, on the Darwinian hypothesis, varies slightly but indefinitely in every organ and every part of every organ, how very generally must favourable

<hr>

[1] " Principles of Biology," vol. i. p. 122.

variations as to the length of the neck have been accompanied by some unfavourable variation in some other part, neutralizing the action of the favourable one, the latter, moreover, only taking effect during these periods of drought! How often must not individuals, favoured by a slightly increased length of neck, have failed to enjoy the elevated foliage which they had not strength or endurance to attain ; while other individuals, exceptionally robust, could struggle on yet further till they arrived at vegetation within their reach.

However, allowing this example to pass, many other instances will be found to present great difficulties.

Let us take the cases of mimicry amongst lepidoptera and other insects. Of this subject Mr. Wallace has given a most interesting and complete account,[1] showing in how many and strange instances this superficial resemblance by one creature to some other quite distinct creature acts as a safeguard to the first. One or two instances must here suffice. In South America there is a family of butterflies, termed *Heliconidæ*, which are very conspicuously coloured and slow in flight, and yet the individuals abound in prodigious numbers, and take no precautions to conceal themselves, even when at rest, during the night. Mr. Bates (the author of the very interesting work " The Naturalist on the River Amazons," and the discoverer of " Mimicry") found that these conspicuous butterflies had a very strong and disagreeable odour; so much so that any one handling them and squeezing them, as a collector must do, has his fingers stained and so infected by the smell, as to require time and much trouble to remove it.

It is suggested that this unpleasant quality is the cause of the abundance of the Heliconidæ ; Mr. Bates and other observers reporting that they have never seen them attacked by the birds, reptiles, or insects which prey upon other lepidoptera.

Now it is a curious fact that very different South American

---

[1] See " Natural Selection," chap. iii. p. 45.

butterflies put on, as it were, the exact dress of these offensive beauties and mimic them even in their mode of flight.

In explaining the mode of action of this protecting resemblance Mr. Wallace observes :[1] "Tropical insectivorous birds very frequently sit on dead branches of a lofty tree, or on those which overhang forest paths, gazing intently around, and darting off at intervals to seize an insect at a considerable distance, with which they generally return to their station to devour. If a bird began by capturing the slow-flying conspicuous Heliconidæ, and found them always so disagreeable that it could not eat them, it would after a very few trials leave off catching them at all ; and their whole appearance, form, colouring, and mode of flight is so peculiar, that there can be little doubt birds would soon learn to distinguish them at a long distance, and never waste any time in pursuit of them. Under these circumstances, it is evident that any other butterfly of a group which birds were accustomed to devour, would be almost equally well protected by closely resembling a Heliconia externally, as if it acquired also the disagreeable odour ; always supposing that there were only a few of them among a great number of Heliconias."

"The approach in colour and form to the Heliconidæ, however, would be at the first a positive, though perhaps a slight, advantage ; for although at short distances this variety would be easily distinguished and devoured, yet at a longer distance it might be mistaken for one of the uneatable group, and so be passed by and gain another day's life, which might in many cases be sufficient for it to lay a quantity of eggs and leave a numerous progeny, many of which would inherit the peculiarity which had been the safeguard of their parent."

As a complete example of mimicry Mr. Wallace refers to a common Indian butterfly. He says :[2] "But the most wonderful and undoubted case of protective resemblance in a butterfly,

[1] Loc. cit. p. 80.   [2] Ibid. p. 59.

which I have ever seen, is that of the common Indian *Kallima inachis*, and its Malayan ally, *Kallima paralekta*. The upper

LEAF BUTTERFLY IN FLIGHT AND REPOSE.
(*From Mr. Wallace's "Malay Archipelago."*)

surface of these is very striking and showy, as they are of a large size, and are adorned with a broad band of rich orange on a deep bluish ground.   The under side is very variable in colour, so that out of fifty specimens no two can be found exactly alike, but every one of them will be of some shade of ash, or brown, or ochre, such as are found among dead, dry, or decaying leaves. The apex of the upper wings is produced into an acute point, a very common form in the leaves of tropical shrubs and trees, and the lower wings are also produced into a short narrow tail.   Between these two points runs a dark curved line exactly representing the midrib of a leaf, and from this radiate on each side a few oblique lines, which serve to indicate the lateral veins of a leaf.   These marks are more clearly seen on the outer portion of the base of the wings, and on the inner side towards the middle and apex, and it is very curious to observe how the usual marginal and transverse striæ of the group are here modified and strengthened so as to become adapted for an imitation of the venation of a leaf." . . . " But this resemblance, close as it is, would be of little use if the habits of the insect did not accord with it.   If the butterfly sat upon leaves or upon flowers, or opened its wings so as to expose the upper surface, or exposed and moved its head and antennæ as many other butterflies do, its disguise would be of little avail.   We might be sure, however, from the analogy of many other cases, that the habits of the insect are such as still further to aid its deceptive garb ; but we are not obliged to make any such supposition, since I myself had the good fortune to observe scores of *Kallima paralekta*, in Sumatra, and to capture many of them, and can vouch for the accuracy of the following details.   These butterflies frequent dry forests, and fly very swiftly.   They were seen to settle on a flower or a green leaf, but were many times lost sight of in a bush or tree of dead leaves.   On such occasions they were generally searched for in vain, for while gazing intently at the very spot where one had disappeared, it would often suddenly dart out, and again vanish twenty or fifty yards further on. On one or two occasions the insect was detected reposing, and it could then be seen how completely it assimilates itself to the surrounding leaves. It sits on a nearly upright twig, the wings fitting closely back to back, concealing the antennæ and head, which are drawn up between their bases. The little tails of the hind wing touch the branch, and form 'a perfect stalk to the leaf, which is supported in its place by the claws of the middle ·pair of feet, which are slender and inconspicuous. The irregular outline of the wings gives exactly the perspective effect of a shrivelled leaf. We thus have size, colour, form, markings, and habits, all combining together to produce a disguise which may be said to be absolutely perfect; and the protection which it affords is sufficiently indicated by the abundance of the individuals that possess it."

Beetles also imitate bees and wasps, as do some Lepidoptera; and objects the most bizarre and unexpected are simulated, such as dung and drops of dew. Some insects, called bamboo and walking-stick insects, have a most remarkable resemblance to pieces of bamboo, to twigs and branches. Of these latter insects Mr. Wallace says :[1] " Some of these are a foot long and as thick as one's finger, and their whole colouring, form, rugosity, and the arrangement of the head, legs, and antennæ, are such as to render them absolutely identical in appearance with dry sticks. They hang loosely about shrubs in the forest, and have the extraordinary habit of stretching out their legs unsymmetrically, so as to render the deception more complete." Now let us suppose that the ancestors of these various animals were all destitute of the very special protections they at present possess, as on the Darwinian hypothesis we must do. Let it also be conceded that small deviations from the antecedent colouring or form would tend to make some of their ancestors escape destruction by causing them more or less frequently to

[1] Loc. cit. p. 64.

D

be passed over, or mistaken by their persecutors. Yet the deviation must, as the event has shown, in each case be in some definite direction, whether it be towards some other animal or plant, or towards some dead or inorganic matter. But as, according to Mr. Darwin's theory, there is a constant tendency to indefinite variation, and as the minute incipient variations will be in *all directions*, they must tend to neutralize each other, and at first to form such unstable modifications that it is difficult, if not impossible, to see how such indefinite oscillations of infinitesimal beginnings can ever build up a sufficiently appreciable resemblance to a leaf, bamboo, or other object, for " Natural Selection " to seize upon and perpetuate. This difficulty is augmented when we consider—a point to be dwelt upon hereafter—how necessary it is that many individuals should be similarly modified simultaneously. This has been insisted on in an able article in the *North British Review* for June 1867, p. 286, and· the consideration of the article has occasioned Mr. Darwin to make an important modification in his views.[1]

In these cases of mimicry it seems difficult indeed to imagine a reason why variations tending in an *infinitesimal degree* in any special direction should be preserved. All variations would be preserved which tended to obscure the perception of an animal by its enemies, whatever direction those variations might take, and the common preservation of conflicting tendencies would greatly favour their mutual neutralization and obliteration if we may rely on the many cases recently brought forward by Mr. Darwin with regard to domestic animals.

Mr. Darwin explains the imitation of some species by others more or less nearly allied to it, by the common origin of both the mimic and the mimicked species, and the consequent possession by both (according to the theory of " Pangenesis ") of gemmules tending to reproduce ancestral characters, which characters the

---

[1] " Origin of Species," 5th edit. p. 104.

mimic must be assumed first to have lost and then to have recovered. Mr. Darwin says,[1] " Varieties of one species frequently mimic distinct species, a fact in perfect harmony with the foregoing cases, and explicable *only on the theory of descent.*" But this at the best is but a partial and very incomplete expla-

THE WALKING-LEAF INSECT.

nation. It is one, moreover, which Mr. Wallace does not accept.[2] It is very incomplete, because it has no bearing on some of the most striking cases, and of course Mr. Darwin does not pretend that it has. We should have to go back far indeed to reach the common ancestor of the mimicking walking-leaf

" Animals and Plants under Domestication," vol. ii. p. 351.

[2] Loc. cit. pp. 109, 110.

D 2

insect and the real leaf it mimics, or the original progenitor of both the bamboo insect and the bamboo itself. As these last most remarkable cases have certainly nothing to do with heredity,[1] it is unwarrantable to make use of that explanation for other protective resemblances, seeing that its inapplicability, in certain instances, is so manifest.

Again, at the other end of the process it is as difficult to account for the last touches of perfection in the mimicry. Some insects which imitate leaves extend the imitation even to the very injuries on those leaves made by the attacks of insects or of fungi. Thus, speaking of one of the walking-stick insects, Mr. Wallace says :[2] " One of these creatures obtained by myself in Borneo (*Ceroxylus laceratus*) was covered over with foliaceous excrescences of a clear olive-green colour, so as exactly to resemble a stick grown over by a creeping moss or jungermannia. The Dyak who brought it me assured me it was grown over with moss although alive, and it was only after a most minute examination that I could convince myself it was not so." Again, as to the leaf butterfly, he says :[3] " We come to a still more extraordinary part of the imitation, for we find representations of leaves in every stage of decay, variously blotched, and mildewed, and pierced with holes, and in many cases irregularly covered with powdery black dots, gathered into patches and spots, so closely resembling the various kinds of minute fungi that grow on dead leaves, that it is impossible to avoid thinking at first sight that the butterflies themselves have been attacked by real fungi."

Here imitation has attained a development which seems utterly beyond the power of the mere " survival of the fittest " to produce. How this double mimicry can importantly aid in the struggle for life seems puzzling indeed, but much more so

---

[1] Heredity is the term used to denote the tendency which there is in offspring to reproduce parental features.

[2] Loc. cit. p. 64.

[3] Loc. cit. p. 60.

how the first faint beginnings of the imitation of such injuries in the leaf can be developed in the animal into such a complete representation of them—*a fortiori* how simultaneous and similar first beginnings of imitations of such injuries could ever have been developed in several individuals, out of utterly indifferent and indeterminate infinitesimal variations in all conceivable directions.

Another instance which may be cited is the asymmetrical condition of the heads of the flat-fishes (Pleuronectidæ), such as the sole, the flounder, the brill, the turbot, &c. In all these

PLEURONECTIDÆ, WITH THE PECULIARLY PLACED EYE IN DIFFERENT POSITIONS.
(*From Dr. Traquair's paper in the "Transactions of the Linnean Society, 1865."*)

fishes the two eyes, which in the young are situated as usual one on each side, come to be placed, in the adult, both on the same side of the head. If this condition had appeared at once, if in the hypothetically fortunate common ancestor of these fishes an eye had suddenly become thus transferred, then the perpetuation of such a transformation by the action of "Natural Selection" is conceivable enough. Such sudden changes, however, are not those favoured by the Darwinian theory, and indeed the accidental occurrence of such a spontaneous transformation is hardly conceivable. But if this is not so, if the transit was

gradual, then how such transit of one eye a minute fraction of
the journey towards the other side of the head could benefit
the individual is indeed far from clear. It seems, even, that
such an incipient transformation must rather have been injurious. Another point with regard to these flat-fishes is that
they appear to be in all probability of recent origin—*i.e.* geolo-
gically speaking. There is, of course, no great stress to be laid
on the mere absence of their remains from the secondary strata,
nevertheless that absence is noteworthy, seeing that existing
fish families, *e.g.* sharks (Squalidæ), have been found abundantly
even down so far as the carboniferous rocks, and traces of them
in the Upper Silurian.

Another difficulty seems to be the first formation of the limbs
of the higher animals. The lowest Vertebrata [1] are perfectly
limbless, and if, as most Darwinians would probably assume, the
primeval vertebrate creature was also apodal, how are the preservation and development of the first rudiments of limbs to be
accounted for—such rudiments being, on the hypothesis in question, infinitesimal and functionless?

In reply to this it has been suggested that a mere flattening
of the end of the body has been useful, such, *e.g.*, as we see in
sea-snakes, [2] which may be the rudiment of a tail formed strictly
to aid in swimming. Also that a mere *roughness* of the skin
might be useful to a swimming animal by holding the water
better, that thus minute processes might be selected and preserved, and that, in the same way, these might be gradually increased into limbs. But it is, to say the least, very questionable
whether a roughness of the skin, or minute processes, would be

---

[1] The term "Vertebrata" denotes that large group of animals which are
characterized by the possession of a spinal column, commonly known as
the "backbone." Such animals are ourselves, together with all beasts,
birds, reptiles, frogs, toads, and efts, and also fishes.

[2] It is hardly necessary to observe that these "sea-snakes" have no
relation to the often-talked-of "sea-serpent." They are small, venomous
reptiles, which abound in the Indian seas.

useful to a swimming animal ; the motion of which they would
as much impede as aid, unless they were at once capable of a
suitable and appropriate action, which is against the hypothesis.
Again, the change from mere indefinite and accidental processes
to two regular pairs of symmetrical limbs, as the result of
merely fortuitous, favouring variations, is a step the feasibility
of which hardly commends itself to the reason, seeing the very
different positions assumed by the ventral fins in different fishes.
If the above suggestion made in opposition to the views here
asserted be true, then the general constancy of position of the
limbs of vertebrata may be considered as due to the position
assumed by the primitive rugosities from which those limbs were
generated.   Clearly only two pairs of rugosities were so preserved
and developed, and all limbs (on this view) are descendants of
the same two pairs, as all have so similar a fundamental structure.   Yet we find in many fishes the pair of fins, which correspond to the hinder limbs of other animals, placed so far
forwards as to be either on the same level with, or actually in
front of, the normally anterior pair of limbs ; and such fishes
are from this circumstance called " thoracic," or "jugular" fishes
respectively, as the weaver fishes and the cod.   This is a
wonderful contrast to the fixity of position of vertebrate limbs
generally.   If then such a change can have taken place in the
comparatively short time occupied by the evolution of these
special fish forms, we might certainly expect other and far
more bizarre structures would (did not some law forbid) have
been developed, from other rugosities, in the manifold exi-
gencies of the multitudinous organisms which must (on the
Darwinian hypothesis) have been gradually evolved during
the enormous period intervening between the first appearance
of vertebrate life and the present day.   Yet, with these exceptions, the position of the limbs is constant from the lower
fishes up to man, there being always an anterior pectoral pair
placed in front of a posterior or pelvic pair when both are

present, and in no single instance are there more than these two pairs.

The development of whalebone (baleen) in the mouth of the whale is another difficulty. A whale's mouth is furnished with very numerous horny plates, which hang down from the palate along each side of the mouth. They thus form two longitudinal series, each plate of which is placed transversely to the long

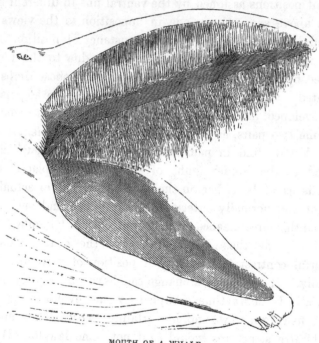

MOUTH OF A WHALE.

axis of the body, and all are very close together. On depressing the lower lip the free outer edges of these plates come into view. Their inner edges are furnished with numerous coarse hair-like processes, consisting of some of the constituent fibres of the horny plates—which, as it were, fray out—and the mouth is thus lined, except below, by a network of countless fibres formed by the inner edges of the two series of plates. This network acts as a sort of sieve. When the whale feeds it takes into its mouth a great gulp of water, which it drives out again through the intervals of the horny plates of baleen, the fluid thus traversing the sieve of horny fibres, which retains the minute creatures on which these marine monsters subsist. Now it is obvious, that if this baleen had once attained such a size and development as to be at all useful, then its preservation and augmentation within serviceable limits, would be promoted by "Natural Selection" alone. But how to obtain the beginning of such useful development? There are indeed certain animals of exclusively aquatic habits (the dugong and manatee) which also possess more or less horn on the palate, and at first sight this might be taken as a mitigation of the difficulty; but it is not so, and the fact does not help us one step further along the road: for, in the first place, these latter animals differ so importantly in structure from whales and porpoises that they form an altogether distinct order, and cannot be thought to approximate to the whale's progenitors. They are vegetarians, the whales feed on animals; the former never have the ribs articulated in the mode in which they are in some of the latter; the

FOUR PLATES OF BALEEN SEEN OBLIQUELY FROM WITHIN.

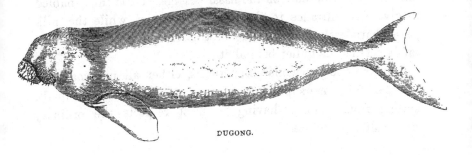

DUGONG.

former have pectoral mammæ, and the latter are provided with two inguinal mammary glands, and have the nostrils enlarged into blowers, which the former have not. The former thus constitute the order Sirenia, while the latter belong to the Cetacea. In the second place, the horny matter on the palates of the dugong and manatee has not, even initially, that "strainer" action which is the characteristic function of the Cetacean "baleen."

There is another very curious structure, the origin or the disappearance of which it seems impossible to account for on the hypothesis of minute indefinite variations. It is that of the mouth of the young kangaroo. In all mammals, as in ourselves, the air-passage from the lungs opens in the floor of the mouth behind the tongue, and in front of the opening of the gullet, so that each particle of food as it is swallowed passes over the opening, but is prevented from falling into it (and thus causing death from choking) by the action of a small cartilaginous shield (the epiglottis), which at the right moment bends back and protects the orifice. Now the kangaroo is born in such an exceedingly imperfect and undeveloped condition, that it is quite unable to suck. The mother therefore places the minute blind and naked young upon the nipple, and then injects milk into it by means of a special muscular envelope of the mammary gland. Did no special provision exist, the young one must infallibly be choked by the intrusion of the milk into the windpipe. But there *is* a special provision. The larynx is so elongated that it rises up into the posterior end of the nasal passage, and is thus enabled to give free entrance to the air for the lungs, while the milk passes harmlessly on each side of this elongated larynx, and so safely attains the gullet behind it.

Now, on the Darwinian hypothesis, either all mammals descended from marsupial progenitors, or else the marsupials, sprung from animals having in most respects the ordinary mammalian structure.

On the first alternative, how did " Natural Selection" remove this (at least perfectly innocent and harmless) structure in almost all other mammals, and, having done so, again reproduce it in precisely those forms which alone require it, namely, the Cetacea ?   That such a harmless structure *need not* be removed any Darwinian must confess, since a structure exists in both the crocodiles and gavials, which enables the former to breathe themselves while drowning the prey which they hold in their mouths.   On Mr. Darwin's hypothesis it could only have been developed where useful, therefore not in the gavials (!) which feed on fish, but which yet retain, as we might expect, this, in them superfluous but harmless formation.

On the second alternative, how did the elongated larynx itself arise, seeing that if its development lagged behind that of the maternal structure, the young primeval kangaroo must be choked: while without the injecting power in the mother, it must be starved ?   The struggle by the sole action of which such a form was developed must indeed have been severe !

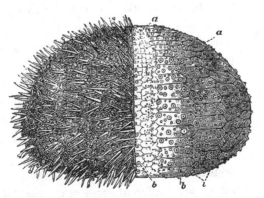

AN ECHINUS, OR SEA-URCHIN
(The spines removed from one-half.)

The sea-urchins (Echinus) present us also with structures the origin of which it seems impossible to explain by the action of "Natural Selection" only. These lowly animals belong to that group of the star-fish class (Echinodermata), the species of which possess generally spheroidal bodies, built up of multitudinous calcareous plates, and constitute the order Echinoidea. They are also popularly known as sea-eggs. Utterly devoid of limbs, the locomotion of these creatures is effected by means of rows of small tubular suckers (which protrude through pores in the calcareous plates) and by moveable spines scattered over the body.

Besides these spines and suckers there are certain very peculiar structures, termed " Pedicellariæ." Each of these consists of a long slender stalk, ending in three short limbs—or rather jaws—the whole supported by a delicate internal skeleton. The three limbs (or jaws), which start from a common point at the end of the stalk, are in the constant habit of opening and closing together again with a snapping action, while the stalk itself sways about. The utility of these appendages is, even now, problematical. It may be that they remove from the surface of the animal's body foreign substances which would be prejudicial to it, and which it cannot otherwise get rid of. But granting this, what would be the utility of the *first rudimentary beginnings* of such structures, and how could such incipient buddings have ever preserved the life of a single Echinus? It is true that on Darwinian principles the ancestral form from which the sea-urchin developed was different, and must not be conceived merely as an Echinus devoid of pedicellariæ; but this makes the difficulty none the less. It is equally hard to imagine that the first rudiments of such structures could have been useful to *any* animal

PEDICELLARIÆ.
(Immensely
enlarged.)

ments of such structures could have been useful to *any* animal from which the Echinus might have been derived. Moreover, not even the *sudden* development of the snapping action could have been beneficial without the freely moveable stalk, nor could the latter have been efficient without the snapping jaws, yet no minute merely indefinite variations could simultaneously evolve these complex co-ordinations of structure; to deny this seems to do no less than to affirm a startling paradox.

Mr. Darwin explains the appearance of some structures, the utility of which is not apparent, by the existence of certain " laws of correlation." By these he means that certain parts or organs of the body are so related to other organs or parts, that when the first are modified by the action of " Natural Selection," or what not, the second are simultaneously affected, and increase proportionally or possibly so decrease. Examples of such are the hair and teeth in the naked Turkish dog, the general deafness of white cats with blue eyes, the relation between the presence of more or less down on young birds when first hatched, and the future colour of their plumage,[1] with many others. But the idea that the modification of any internal or external part of the body of an Echinus carries with it the effect of producing elongated, flexible, triradiate, snapping processes, is, to say the very least, fully as obscure and mysterious as what is here contended for, viz. the efficient presence of an unknown internal natural law or laws conditioning the evolution of new specific forms from preceding ones, modified by the action of surrounding conditions, by " Natural Selection" and by other controlling influences.

The same difficulty seems to present itself in other examples of exceptional structure and action. In the same Echinus, as in many allied forms, and also in some more or less remote ones, a very peculiar mode of development exists. The adult is not

---

[1] " Origin of Species," 5th edit., 1869, p. 179.

formed from the egg directly, but the egg gives rise to a creature which swims freely about, feeds, and is even somewhat complexly organized. Soon a small lump appears on one side of its stomach ; this enlarges, and, having established a communication with the exterior, envelopes and appropriates the creature's stomach, with which it swims away and develops into the complete adult form, while the dispossessed individual perishes.

Again, certain flies present a mode of development equally bizarre, though quite different. In these flies, the grub is, as usual, produced from the ovum, but this grub, instead of growing up into the adult in the ordinary way, undergoes a sort of liquefaction of a great part of its body, while certain patches of formative tissue, which are attached to the ramifying air tubes, or tracheæ (and which patches bear the name of "imaginal disks"), give rise to the legs, wings, eyes, &c., respectively ; and these severally formed parts grow together, and build up the head and body by their mutual approximation. Such a process is unknown outside the class of insects, and inside that class it is only known in a few of the two-winged flies. Now, how "Natural Selection," or any "laws of correlation," can account for the gradual development of such an exceptional process of development—so extremely divergent from that of other insects—seems nothing less than inconceivable. Mr. Darwin himself[1] gives an account of a very peculiar and abnormal mode of development of a certain beetle, the sitaris, as described by M. Fabre. This insect, instead of at first appearing in its grub stage, and then, after a time, putting on the adult form, is at first active and furnished with six legs, two long antennæ, and four eyes. Hatched in the nests of bees, it at first attaches itself to one of the males, and then crawls, when the opportunity offers, upon a female bee. When the female bee lays her eggs, the young sitaris springs upon them

[1] "Origin of Species," 5th edit., p. 532.

and devours them. Then, losing its eyes, legs, and antennæ, and becoming rudimentary, it sinks into an ordinary grub-like form, and feeds on honey, ultimately undergoing another transformation, re-acquiring its legs, &c., and emerging a perfect beetle! That such a process should have arisen by the accumulation of minute accidental variations in structure and habit, appears to many minds, quite competent to form an opinion on the subject, absolutely incredible.

It may be objected, perhaps, that these difficulties are *difficulties of ignorance*—that we cannot explain them because we do not know *enough* of the animals. But it is here contended that this is not the case ; it is not that we merely fail to see how Natural Selection acted, but that there is a positive incompatibility between the cause assigned and the results. It will be stated shortly what wonderful instances of co-ordination and of unexpected utility Mr. Darwin has discovered in orchids. The discoveries are not disputed or undervalued, but the explanation of their *origin* is deemed thoroughly unsatisfactory—utterly insufficient to explain the incipient, infinitesimal beginnings of structures which are of utility only when they are considerably developed.

Let us consider the mammary gland, or breast. Is it conceivable that the young of any animal was ever saved from destruction by accidentally sucking a drop of scarcely nutritious fluid from an accidentally hypertrophied cutaneous gland of its mother ? And even if one was so, what chance was there of the perpetuation of such a variation ? On the hypothesis of Natural Selection itself, we must assume that up to that time the race had been well adapted to the surrounding conditions ; the temporary and accidental trial and change of conditions, which caused the so-sucking young one to be the " fittest to survive " under the supposed circumstances, would soon cease to act, and then the progeny of the mother, with the accidentally hypertrophied, sebaceous glands, would have no tendency to survive the far outnumbering descendants of the normal ancestral form. If, on the other hand, we assume the change of conditions not to have been temporary but permanent, and also assume that this permanent change of conditions was accidentally synchronous with the change of structure, we have a coincidence of very remote probability indeed. But if, again, we accept the presence of some harmonizing law simultaneously determining the two changes, or connecting the second with the first by causation, then, of course, we remove the accidental character of the coincidence.

Again, how explain the external position of the male sexual glands in certain mammals? The utility of the modification, when accomplished, is problematical enough, and no less so the incipient stages of the descent.

As was said in the first chapter, Mr. Darwin explains the brilliant plumage of the peacock or the humming-bird by the action of sexual selection : the more and more brilliant males being selected by the females (which are thus attracted) to become the fathers of the next generation, to which generation they tend to communicate their own bright nuptial vesture. But there are peculiarities of colour and of form which it is exceedingly difficult to account for by any such action. Thus, amongst apes, the female is notoriously weaker, and is armed with much less powerful canine tusks than the male. When we consider what is known of the emotional nature of these animals, and the periodicity of its intensification, it is hardly credible that a female would often risk life or limb through her admiration of a trifling shade of colour, or an infinitesimally greater though irresistibly fascinating degree of wartiness.[1]

[1] Mr. A. D. Bartlett, of the Zoological Society, informs me that at these periods female apes admit with perfect readiness the access of any males of different species. To be sure this is in confinement ; but the fact is, I think, quite conclusive against any such sexual selection in a state of nature as would account for the local coloration referred to.

Yet the males of some kinds of ape are adorned with quite exceptionally brilliant local decoration, and the male orang is provided with remarkable, projecting, warty lumps of skin upon the cheeks. As we have said, the weaker female can hardly be supposed to have developed these by persevering and long-continued selection, nor can they be thought to tend to the preservation of the individual. On the contrary, the presence of this enlarged appendage must occasion a slight increase in the need of nutriment, and in so far must be a detriment,

RATTLESNAKE.

although its detrimental effect would not be worth speaking of except in relation to " Darwinism," according to which,

E

"selection" has acted through unimaginable ages, and has ever tended to suppress any useless development by the struggle for life.[1]

In poisonous serpents, also, we have structures which, at all events at first sight, seem positively hurtful to those reptiles. Such are the rattle of the rattlesnake, and the expanding neck of the cobra, the former seeming to warn the ear of the intended

COBRA.
(*Copied, by permission, from Sir Andrew Smith's "Reptiles of South Africa."*)

victim, as the latter warns the eye. It is true we cannot perhaps demonstrate that the victims are alarmed and warned, but, on

---

[1] Mr. Darwin, in the last (fifth) edition of "Natural Selection," 1869, p. 102, admits that all sexual differences are not to be attributed to the agency of sexual selection, mentioning the wattle of carrier pigeons, tuft of turkey-cock, &c. These characters, however, seem less inexplicable by sexual selection than those given in the text.

Darwinian principles, they certainly ought to be so. For the rashest and most incautious of the animals preyed on would always tend to fall victims, and the existing individuals being the long-descended progeny of the timid and cautious, ought to have an inherited tendency to distrust, amongst other things, both "rattling" and "expanding" snakes. As to any power of fascination exercised by means of these actions, the most distinguished naturalists, certainly the most distinguished erpetologists, entirely deny it, and it is opposed to the careful observations of those known to us.[1]

The mode of formation of both the eye and the ear of the highest animals is such that, if it is (as most Darwinians assert processes of development to be) a record of the actual steps by which such structures were first evolved in antecedent forms, it almost amounts to a demonstration that those steps were never produced by "Natural Selection."

The eye is formed by a simultaneous and corresponding ingrowth of one part and outgrowth of another. The skin in front of the future eye becomes depressed, the depression increases and assumes the form of a sac, which changes into the aqueous humour and lens. An outgrowth of brain substance, on the other hand, forms the retina, while a third process is a lateral ingrowth of connective tissue, which afterwards changes into the vitreous humour of the eye.

The internal ear is formed by an involution of the integument, and not by an outgrowth of the brain. But tissue, in connexion with it, becomes in part changed, thus forming the auditory nerve, which places the tegumentary sac in direct communication with the brain itself.

---

[1] I am again indebted to the kindness of Mr. A. D. Bartlett, amongst others. That gentleman informs me that, so far from any mental emotion being produced in rabbits by the presence and movements of snakes, that he has actually seen a male and female rabbit satisfy the sexual instinct in that presence, a rabbit being seized by a snake when *in coitu.*

Now, these complex and simultaneous co-ordinations could never have been produced by infinitesimal beginnings, since, until so far developed as to effect the requisite junctions, they are useless. But the eye and ear when fully developed present conditions which are hopelessly difficult to reconcile with the mere action of "Natural Selection." The difficulties with regard to the eye have been well put by Mr. Murphy, especially that of the concordant result of visual development springing from different starting-points and continued on by independent roads.

He says,[1] speaking of the beautiful structure of the perfect eye, "The higher the organization, whether of an entire organism or of a single organ, the greater is the number of the parts that co-operate, and the more perfect is their co-operation ; and consequently, the more necessity there is for corresponding variations to take place in all the co-operating parts at once, and the more useless will be any variation whatever unless it is accompanied by corresponding variations in the co-operating parts ; while it is obvious that the greater the number of variations which are needed in order to effect an improvement, the less will be the probability of their all occurring at once. It is no reply to this to say, what is no doubt abstractedly true, that whatever is possible becomes probable, if only time enough be allowed. There are improbabilities so great that the common sense of mankind treats them as impossibilities. It is not, for instance, in the strictest sense of the word, impossible that a poem and a mathematical proposition should be obtained by the process of shaking letters out of a box ; but it is improbable to a degree that cannot be distinguished from impossibility ; and the improbability of obtaining an improvement in an organ by means of several spontaneous variations, all occurring together, is an improbability of the same kind. If we suppose that any single variation occurs

[1] "Habit and Intelligence," vol. i. p. 319.

on the average once in m times, the probability of that variation occurring in any individual will be

$$\frac{1}{m};$$

and suppose that x variations must concur in order to make an
improvement, then the probability of the necessary variations
all occurring together will be

$$\frac{1}{m^x}.$$

Now suppose, what I think a moderate proposition, that the
value of m is 1,000, and the value of x is 10, then

$$\frac{1}{m^x} = \frac{1}{1000^{10}} = \frac{1}{10^{30}}.$$

A number about ten thousand times as great as the number of
waves of light that have fallen on the earth since historical
time began.   And it is to be further observed, that no improvement will give its possessor a *certainty* of surviving and leaving
offspring, but only an *extra chance*, the value of which it is
quite impossible to estimate." This difficulty is, as Mr. Murphy
points out, greatly intensified by the undoubted fact that the
wonderfully complex structure has been arrived at quite independently in beasts on the one hand and in cuttle-fishes on the
other; while creatures of the insect and crab division present
us with a third and quite separately developed complexity.

As to the ear, it would take up too much space to describe its
internal structure;[1] it must suffice to say that in its interior there
is an immense series of minute rod-like bodies, termed *fibres of
Corti,* having the appearance of a key-board, and each fibre being
connected with a filament of the auditory nerve, these nerves
being like strings to be struck by the keys, *i.e.* by the fibres of
Corti.   Moreover, this apparatus is supposed to be a key-board

[1] The reader may consult Huxley's "Lessons in Elementary Physiology," p. 204.

in function as well as in appearance, the vibration of each one fibre giving rise, it is believed, to the sensation of one particular tone, and combinations of such vibrations producing chords. It is by the action of this complex organ then, that all the wonderful intricacy and beauty of Beethoven and Mozart come, most probably, to be perceived and appreciated.

Now it can hardly be contended that the preservation of any race of men in the struggle for life ever depended on such an extreme delicacy and refinement of the internal ear,—a perfection only exercised in the enjoyment and appreciation of the most perfect musical performances. How, then, could either the minute incipient stages, or the final perfecting touches of this admirable structure, have been brought about by vague, aimless, and indefinite variations in all conceivable directions of an organ, suitable to enable the rudest savage to minister to his necessities, but no more?

Mr. Wallace[1] makes an analogous remark with regard to the organ of voice in man—the human larynx. He says of singing: "The habits of savages give no indication of how this faculty could have been developed by Natural Selection, because it is never required or used by them. The singing of savages is a more or less monotonous howling, and the females seldom sing at all. Savages certainly never choose their wives for fine voices, but for rude health, and strength, and physical beauty. Sexual selection could not therefore have developed this wonderful power, which only comes into play among civilized people."

Reverting once more to beauty of form and colour, there is one manifestation of it for which no one can pretend that sexual selection can possibly account. The instance referred to is that presented by bivalve shell-fish.[2] Here we meet with charming tints and elegant forms and markings of no direct use to their

---

[1] "Natural Selection," p. 350.

[2] Bivalve shell-fish are creatures belonging to the oyster, scallop, and cockle group, *i.e.* to the class Lamellibranchiata.

possessors [1] in the struggle for life, and of no indirect utility as regards sexual selection, for fertilization takes place by the mere action of currents of water, and the least beautiful individual has fully as good a chance of becoming a parent as has the one which is the most favoured in beauty of form and colour.

Again, the peculiar outline and coloration of certain orchids —notably of our own bee, fly, and spider orchids—seem hardly explicable by any action of "Natural Selection." Mr. Darwin says very little on this singular resemblance of flowers to insects, and what he does say seems hardly to be what an advocate of "Natural Selection" would require. Surely, for minute accidental indefinite variations to have built up such a striking resemblance to insects, we ought to find that the preservation of the plant, or the perpetuation of its race, depends almost constantly on relations between bees, spiders, and flies respectively and the bee, spider, and fly orchids.[2] This process must have continued for ages constantly and perseveringly, and yet what is the fact? Mr. Darwin tells us, in his work on the Fertilization of Orchids, that neither the spider nor the fly orchids are much visited by insects, while, with regard to the bee orchid, he says, " I have never seen an insect visit these flowers." And he shows how this species is even wonderfully and specially modified to effect self-fertilization.

In the work just referred to Mr. Darwin gives a series of the most wonderful and minute contrivances by which the visits of insects are utilized for the fertilization of orchids,—structures

---

[1] The attempt has been made to explain these facts as owing to "manner and symmetry of growth, and to colour being incidental on the chemical nature of the constituents of the shell." But surely beauty depends on some such matters in *all* cases !

[2] It has been suggested in opposition to what is here said, that there is no real resemblance, but that the likeness is *"fanciful!"* The denial, however, of the fact of a resemblance which has struck so many observers, reminds one of the French philosopher's estimate of facts hostile to his theory—" Tant pis pour les faits !"

so wonderful that nothing could well be more so, except the attribution of their origin to minute, fortuitous, and indefinite variation.

The instances are too numerous and too long to quote, but in his " Origin of Species " [1] he describes two which must not be passed over. In one (*Coryanthes*) the orchid has its lower lip enlarged into a bucket, above which stand two water-secreting horns. These latter replenish the bucket from which, when half-filled, the water overflows by a spout on one side. Bees visiting the flower fall into the bucket and crawl out at the spout. By the peculiar arrangement of the parts of the flower, the first bee which does so carries away the pollen-mass glued to his back, and then when he has his next involuntary bath in another flower, as he crawls out the pollen-mass attached to him comes in contact with the stigma of that second flower and fertilizes it. In the other example (*Catasetum*), when a bee gnaws a certain part of the flower, he inevitably touches a long delicate projection, which Mr. Darwin calls the antenna. " This antenna transmits a vibration to a certain membrane, which is instantly ruptured; this sets free a spring by which the pollen-mass is shot forth like an arrow in the right direction, and adheres by its viscid extremity to the back of the bee !"

Another difficulty, and one of some importance, is presented by those communities of ants which have not only a population of sterile females, or workers, but two distinct and very different castes of such. Mr. Darwin believes that he has got over this difficulty by having found individuals intermediate in form and structure [2] between the two working castes; others may think

---

[1] Fifth Edition, p. 236.

[2] Mr. Smith, of the Entomological department of the British Museum, has kindly informed me that the individuals intermediate in structure are very few in number—not more than five per cent.—compared with the number of distinctly differentiated individuals. Besides, in the Brazilian kinds these intermediate forms are wanting.

that we have in this belief of Mr. Darwin, an example of the unconscious action of volition upon credence. A vast number of difficulties similar to those which have been mentioned might easily be cited—those given, however, may suffice.

There remains, however, to be noticed a very important consideration, which was brought forward in the *North British Review* for June 1867, p. 286, namely, the necessity for the simultaneous modification of *many individuals.* This consideration seems to have escaped Mr. Darwin, for at p. 104 of his last (fifth) edition of "Natural Selection," he admits, with great candour, that until reading this article he did not "appreciate how rarely single variations, whether slight or strongly marked, could be perpetuated."

The *North British Review* (speaking of the supposition that a species is changed by the survival of a few individuals in a century through a similar and favourable variation) says : "It is very difficult to see how this can be accomplished, even when the variation is eminently favourable indeed ; and still more difficult when the advantage gained is very slight, as must generally be the case. The advantage, whatever it may be, is utterly outbalanced by numerical inferiority. A million creatures are born ; ten thousand survive to produce offspring. One of the million has twice as good a chance as any other of surviving ; but the chances are fifty to one against the gifted individuals being one of the hundred survivors. No doubt the chances are twice as great against any one other individual, but this does not prevent their being enormously in favour of *some* average individual. However slight the advantage may be, if it is shared by half the individuals produced, it will probably be present in at least fifty-one of the survivors, and in a larger proportion of their offspring ; but the chances are against the preservation of any one 'sport' (*i.e.* sudden, marked variation) in a numerous tribe. The vague use of an imperfectly understood doctrine of ·chance has led Darwinian supporters, first, to confuse the two

cases above distinguished; and, secondly, to imagine that a
very slight balance in favour of some individual sport must
lead to its perpetuation. All that can be said is that in the
above example the favoured sport would be preserved once in
fifty times. Let us consider what will be its influence on the
main stock when preserved. It will breed and have a progeny
of say 100; now this progeny will, on the whole, be intermediate between the average individual and the sport. The odds
in favour of one of this generation of the new breed will be,
say one and a half to one, as compared with the average individual; the odds in their favour will, therefore, be less than
that of their parents; but owing to their greater number, the
chances are that about one and a half of them would survive.
Unless these breed together, a most improbable event, their
progeny would again approach the average individual; there
would be 150 of them, and their superiority would be, say in
the ratio of one and a quarter to one; the probability would
now be that nearly two of them would survive, and have 200
children, with an eighth superiority. Rather more than two
of these would survive; but the superiority would again dwindle,
until after a few generations it would no longer be observed,
and would count for no more in the struggle for life than any
of the hundred trifling advantages which occur in the ordinary
organs. An illustration will bring this conception home. Suppose a white man to have been wrecked on an island inhabited
by negroes, and to have established himself in friendly relations
with a powerful tribe, whose customs he has learnt. Suppose
him to possess the physical strength, energy, and ability of a
dominant white race, and let the food and climate of the island
suit his constitution; grant him every advantage which we can
conceive a white to possess over the native; concede that in
the struggle for existence his chance of a long life will be
much superior to that of the native chiefs; yet from all these
admissions, there does not follow the conclusion that, after a

limited or unlimited number of generations, the inhabitants of
the island will be white.   Our shipwrecked hero would probably
become king; he would kill a great many blacks in the struggle
for existence; he would have a great many wives and children." . . . . "In the first generation there will be some dozens
of intelligent young mulattoes, much superior in average intelligence to the negroes.   We might expect the throne for some
generations to be occupied by a more or less yellow king; but can
any one believe that the whole island will gradually acquire a
white, or even a yellow, population ?"

"Darwin says that in the struggle for life a grain may turn
the balance in favour of a given structure, which will then be
preserved.   But one of the weights in the scale of nature is due
to the number of a given tribe.   Let there be 7000 A's and
7000 B's, representing two varieties of a given animal, and let
all the B's, in virtue of a slight difference of structure, have
the better chance of life by $\frac{1}{7000}$ part.   We must allow that
there is a slight probability that the descendants of B will sup-
plant the descendants of A; but let there be only 7001 A's against
7000 B's at first, and the chances are once more equal, while
if there be 7002 A's to start, the odds would be laid on the A's.
True, they stand a greater chance of being killed; but then they
can better afford to be killed.   The grain will only turn the
scales when these are very nicely balanced, and an advantage
in numbers counts for weight, even as an advantage in structure.   As the numbers of the favoured variety diminish, so
must its relative advantages increase, if the chance of its existence is to surpass the chance of its extinction, until hardly
any conceivable advantage would enable the descendants of a
single pair to exterminate the descendants of many thousands
if they and their descendants are supposed to breed freely with
the inferior variety, and so gradually lose their ascendency."

Mr. Darwin himself says of the article quoted : "The justice
of these remarks cannot, I think, be disputed.   If, for instance,

a bird of some kind could procure its food more easily by
having its beak curved, and if one were born with its beak
strongly curved, and which consequently flourished, nevertheless there would be a very poor chance of this one individual
perpetuating its kind to the exclusion of the common form."
This admission seems almost to amount to a change of front
in the face of the enemy !

These remarks have been quoted at length because they so
greatly intensify the difficulties brought forward in this chapter.
If the most favourable variations have to contend with such
difficulties, what must be thought as to the chance of preservation of the slightly displaced eye in a sole or of the incipient
development of baleen in a whale?

SUMMARY AND CONCLUSION.

It has been here contended that a certain few facts, out of
many which might have been brought forward, are inconsistent with the origination of species by "Natural Selection"
only or mainly.

Mr. Darwin's theory requires minute, indefinite, fortuitous
variations of all parts in all directions, and he insists that
the sole operation of "Natural Selection" upon such is sufficient to account for the great majority of organic forms, with
their most complicated structures, intricate mutual adaptations
and delicate adjustments.

To this conception has been opposed the difficulties presented
by such a structure as the form of the giraffe, which ought not
to have been the solitary structure it is ; also the minute beginnings and the last refinements of protective mimicry equally
difficult or rather impossible to account for by "Natural Selection." Again the difficulty as to the heads of flat-fishes has
been insisted on, as also the origin, and at the same time the
constancy, of the limbs of the highest animals. Reference has
also been made to the whalebone of whales, and to the im-

possibility of understanding its origin through "Natural Selection" only; the same as regards the infant kangaroo, with its singular deficiency of power compensated for by maternal structures on the one hand, to which its own breathing organs bear direct relation on the other. Again, the delicate and complex pedicellariæ of Echinoderms, with a certain process of development (through a secondary larva) found in that class, together with certain other exceptional modes of development, have been brought forward. The development of colour in certain apes, the hood of the cobra, and the rattle of the rattlesnake have also been cited. Again, difficulties as to the process of formation of the eye and ear, and as to the fully developed condition of those complex organs, as well as of the voice, have been considered. The beauty of certain shell-fish; the wonderful adaptations of structure, and variety of form and resemblance, found in orchids; together with the complex habits and social conditions of certain ants, have been hastily passed in review. When all these complications are duly weighed and considered, and when it is borne in mind how necessary it is for the permanence of a new variety that many individuals in each case should be simultaneously modified, the cumulative argument seems irresistible.

The Author of this book can say that though by no means disposed originally to dissent from the theory of "Natural Selection," if only its difficulties could be solved, he has found each successive year that deeper consideration and more careful examination have more and more brought home to him the inadequacy of Mr. Darwin's theory to account for the preservation and intensification of incipient, specific, and generic characters. That minute, fortuitous, and indefinite variations could have brought about such special forms and modifications as have been enumerated in this chapter, seems to contradict not imagination, but reason.

That either many individuals amongst a species of butterfly should be simultaneously preserved through a similar accidental and minute variation in one definite direction, when variations in many other directions would also preserve; or that one or two so varying should succeed in supplanting the progeny of thousands of other individuals, and that this should by no other cause be carried so far as to produce the appearance (as we have before stated) of spots of fungi, &c.—are alternatives of an improbability so extreme as to be practically equal to impossibility.

In spite of all the resources of a fertile imagination, the Darwinian, pure and simple, is reduced to the assertion of a paradox as great as any he opposes. In the place of a mere assertion of our ignorance as to the way these phenomena have been produced, he brings forward, as their explanation, a cause which it is contended in this work is demonstrably insufficient.

Of course in this matter, as elsewhere throughout nature, we have to do with the operation of fixed and constant natural laws, and the knowledge of these may before long be obtained by human patience or human genius; but there is, it is believed, already enough evidence to show that these as yet unknown natural laws or law will never be resolved into the action of " Natural Selection," but will constitute or exemplify a mode and condition of organic action of which the Darwinian theory takes no account whatsoever.

CHAPTER III.

THE CO-EXISTENCE OF CLOSELY SIMILAR STRUCTURES OF
DIVERSE ORIGIN.

Chances against concordant variations.—Examples of discordant ones.—
Concordant variations not unlikely on a non-Darwinian evolutionary
hypothesis.—Placental and implacental mammals.—Birds and reptiles.
—Independent origins of similar sense organs.—The ear.—The eye.—
Other coincidences.—Causes besides Natural Selection produce concor-
dant variations in certain geographical regions.—Causes besides Natural
Selection produce concordant variations in certain zoological and bo-
tanical groups.—There are homologous parts not genetically related.
—Harmony in respect of the organic and inorganic worlds.—Summary
and conclusion.

THE theory of "Natural Selection" supposes that the varied
forms and structure of animals and plants have been built up
merely by indefinite, fortuitous,[1] minute variations in every part
and in all directions—those variations only being preserved which
are directly or indirectly useful to the individual possessing them,
or necessarily correlated with such useful variations.

On this theory the chances are almost infinitely great against
the independent, accidental occurrence and preservation of two
similar series of minute variations resulting in the independent
development of two closely similar forms. In all cases, no
doubt (on this same theory), *some* adaptation to habit or need

---

[1] By accidental variations Mr. Darwin does not, of course, mean to
imply variations really due to "chance," but to utterly indeterminate
antecedents.

would gradually be evolved, but that adaptation would surely be arrived at by different roads. The organic world supplies us with multitudes of examples of similar functional results being attained by the most diverse means. Thus the body is sustained in the air by birds and by bats. In the first case it is so sustained by a limb in which the bones of the hand are excessively reduced, but which is provided with immense outgrowths from the skin—namely, the feathers of the wing. In the second case, however, the body is sustained in the air by a limb in which the

WINGBONES OF PTERODACTYLE, BAT, AND BIRD.
(Copied, by permission, from Mr. Andrew Murray's "Geographical Distribution of Mammals.")

bones of the hand are enormously increased in length, and so sustain a great expanse of naked skin, which is the flying membrane of the bat's wing. Certain fishes and certain reptiles can also flit and take very prolonged jumps in the air. The flying-fish, however, takes these by means of a great elongation of the rays of the pectoral fins—parts which cannot be said to be of the same nature as the constituents of the wing of either the bat or the bird. The little lizard, which enjoys the formidable name of "flying-dragon," flits by means of a structure altogether peculiar—namely, by the liberation and great elongation of some of the ribs which support a fold of skin. In the extinct pterodactyles—

which were *truly* flying reptiles—we meet with an approximation to the structure of the bat, but in the pterodactyle we have only

SKELETON OF THE FLYING-DRAGON.
(Showing the elongated ribs which support the flitting organ.)

one finger elongated in each hand : a striking example of how the very same function may be provided for by a modification similar in principle, yet surely manifesting the independence of its origin. When we go to lower animals, we find flight produced by organs, as the wings of insects, which are not even modified limbs at all; or we find even the function sometimes subserved by quite artificial means, as in the aërial spiders, which use their own threads to float with in the air. In the vegetable kingdom the atmosphere is often made use of for the scattering of seeds, by their being furnished with special structures of very different kinds. The diverse modes by which such seeds are dispersed are well expressed by Mr. Darwin. He says:[1] " Seeds

[1] "Origin of Species," 5th edition, p. 235.

are disseminated by their minuteness,—by their capsule being converted into a light balloon-like envelope,—by being embedded in pulp or flesh, formed of the most diverse parts, and rendered nutritious, as well as conspicuously coloured, so as to attract and be devoured by birds,—by having hooks and grapnels of many kinds and serrated awns, so as to adhere to the fur of quadrupeds,—and by being furnished with wings and plumes, as different in shape as elegant in structure, so as to be wafted by every breeze."

Again, if we consider the poisoning apparatus possessed by different animals, we find in serpents a perforated—or rather very deeply channelled—tooth. In wasps and bees the sting is formed of modified parts, accessory in reproduction. In the scorpion, we have the median terminal process of the body specially organized. In the spider, we have a specially constructed antenna; and finally in the centipede a pair of modified thoracic limbs.

A CENTIPEDE.

It would be easy to produce a multitude of such instances of similar ends being attained by dissimilar means, and it is here contended that by "the action of Natural Selection" *only* it is so improbable as to be practically impossible for two exactly similar structures to have ever been independently developed. It is so because the number of possible variations is indefinitely great, and it is therefore an indefinitely great number to one against a similar series of variations occurring and being similarly preserved in any two independent instances.

The difficulty here asserted applies, however, only to pure Darwinism, which makes use *only* of indirect modifications through the survival of the fittest.

Other theories (for example, that of Mr. Herbert Spencer) admit the *direct* action of conditions upon animals and plants— in ways not yet fully understood—there being conceived to be at the same time a certain peculiar but limited power of response and adaptation in each animal and plant so acted on. Such theories have not to contend against the difficulty proposed, and it is here urged that even very complex extremely similar structures have again and again been developed quite independently one of the other, and this because the process has taken place not by merely haphazard, indefinite variations in all directions, but by the concurrence of some other and internal natural law or laws co-operating with external influences and with Natural Selection in the evolution of organic forms.

It must never be forgotten that to admit any such constant operation of any such unknown natural cause is to deny the purely Darwinian theory, which relies upon the survival of the fittest by means of minute fortuitous indefinite variations.

Amongst many other obligations which the Author has to acknowledge to Professor Huxley, are the pointing out of this very difficulty, and the calling his attention to the striking resemblance between certain teeth of the dog and of the thylacine as one instance, and certain ornithic peculiarities of pterodactyles as another.

Mammals[1] are divisible into one great group, which comprises

[1] *I.e.* warm-blooded animals which suckle their young, such as apes bats, hoofed beasts, lions, dogs, bears, weasels, rats, squirrels, armadillos, sloths, whales, porpoises, kangaroos, opossums, &c.

the immense majority of kinds termed, from their mode of reproduction, *placental Mammals,* and into another very much smaller group comprising the pouched-beasts or marsupials (which are the kangaroos, bandicoots, phalangers, &c., of Australia), and the true opossums of America, called *implacental Mammals.* Now the placental mammals are subdivided into various orders, amongst which are the flesh-eaters (Carnivora, *i.e.* cats, dogs, otters, weasels, &c.), and the insect-eaters (Insectivora, *i.e.* moles, hedgehogs, shrew-mice, &c.). The marsupial mammals also present a variety of forms (some of which are carnivorous beasts, whilst others are insectivorous), so marked that it has been even proposed to divide them into orders parallel to the orders of placental beasts.

The resemblance, indeed, is so striking as, on Darwinian principles, to suggest the probability of genetic affinity; and it even led Professor Huxley, in his Hunterian Lectures, in 1866, to promulgate the notion that a vast and widely-diffused marsupial fauna may have existed anteriorly to the development of the ordinary placental, non-pouched beasts, and that the carnivorous, insectivorous, and herbivorous placentals may have respectively descended from the carnivorous, insectivorous, and herbivorous marsupials.

Amongst other points Professor Huxley called attention to the resemblance between the anterior molars of the placental

TEETH OF UROTRICHUS AND PERAMELES.

dog with those of the marsupial thylacine. These, indeed, are strikingly similar, but there are better examples still of this sort of coincidence. Thus it has often been remarked that the insectivorous marsupials, *e.g. Perameles*, wonderfully correspond, as to the form of certain of the grinding teeth, with certain insectivorous placentals, *e.g. Urotrichus.*

Again, the saltatory insectivores of Africa (*Macroscelides*) not only resemble the kangaroo family (*Macropodidæ*) in their jumping habits and long hind legs, but also in the structure of their molar teeth, and even further, as I have elsewhere[1] pointed out, in a certain similarity of the upper cutting teeth, or incisors.

Now these correspondences are the more striking when we bear in mind that a similar dentition is often put to very different uses. The food of different kinds of apes is very different, yet how uniform is their dental structure! Again, who, looking at the teeth of different kinds of bears, would ever suspect that one kind was frugivorous, and another a devourer exclusively of animal food?

The suggestion made by Professor Huxley was therefore one which had much to recommend it to Darwinians, though it has not met with any notable acceptance, and though he seems himself to have returned to the older notion, namely, that the pouched-beasts, or marsupials, are a special ancient offshoot from the great mammalian class.

But whichever view may be the correct one, we have in either case a number of forms similarly modified in harmony with surrounding conditions, and eloquently proclaiming some natural plastic power, other than mere fortuitous variation with survival of the fittest. If, however, the Reader thinks that teeth are parts peculiarly qualified for rapid variation (in which view the Author cannot concur), he is requested to suspend his judgment till he has considered the question of the independent evolution of the *highest organs of sense.* If this seems to establish the existence

---

[1] " Journal of Anatomy and Physiology" (1868), vol. ii. p. 139.

of some other law than that of "Natural Selection," then the operation of that other law may surely be also traced in the harmonious co-ordinations of dental form.

The other difficulty, kindly suggested to me by the learned Professor, refers to the structure of birds, and of extinct reptiles more or less related to them.

The class of birds is one which is remarkably uniform in its organization. So much is this the case, that the best mode of subdividing the class is a problem of the greatest difficulty. Existing birds, however, present forms which, though closely resembling in the greater part of their structure, yet differ importantly the one from the other. One form is exemplified by the ostrich, rhea, emeu, cassowary, apteryx, dinornis, &c. These are the *struthious* birds. All other existing birds belong to the second division, and are called (from the keel on the breast-bone) *carinate* birds.

Now birds and reptiles have such and so many points in common, that Darwinians must regard the former as modified descendants of ancient reptilian forms. But on Darwinian principles it is impossible that the class of birds so uniform and homogeneous should have had a double reptilian origin. If one set of birds sprang from one set of reptiles, and another set of birds from another set of reptiles, the two sets could never, by "Natural Selection" only, have grown into such a perfect similarity. To admit such a phenomenon would be equivalent to abandoning the theory of "Natural Selection" as the sole origin of species.

Now, until recently it has generally been supposed by evolutionists that those ancient flying reptiles, the pterodactyles, or forms allied to them, were the progenitors of the class of birds; and certain parts of their structure especially support this view. Allusion is here made to the bladebone (scapula), and the bone which passes down from the shoulder-joint to the breast-bone (viz. the coracoid). These bones are such remarkable anticipations of the same parts in ordinary (*i.e.* carinate) birds that it is hardly possible for a Darwinian not to regard the resemblance as due to community of origin.   This resemblance was carefully pointed out by Professor Huxley in his "Hunterian Course" for 1867, when attention was called to the existence in *Dimorphodon macronyx* of even that small process which in birds gives attachment to the upper end of the merrythought.   Also Mr. Seeley [1] has shown that in pterodactyles, as in birds, the optic lobes of the brain were placed low down on each side—"lateral and depressed."   Nevertheless, the view has been put forward and ably maintained by the same Professor,[2] as also by Professor Cope in the United States, that the line of descent from reptiles to birds has not been from ordinary reptiles, through pterodactyle-like forms, to ordinary birds, but to the struthious ones from certain extinct reptiles termed Dinosauria ; one of the most familiarly known of which is the Iguanodon of the Wealden formation.   In these Dinosauria we find skeletal characters unlike those of ordinary (*i.e.* carinate) birds, but closely resembling in certain points the osseous structure of the struthious birds.   Thus a difficulty presents itself as to the explanation of the three following relationships:—(1) That of the Pterodactyles with carinate birds; (2) that of the Dinosauria with struthious birds ; (3) that of the carinate and struthious birds with each other.

Either birds must have had two distinct origins whence they grew to their present conformity, or the very same skeletal, and probably cerebral characters must have spontaneously and independently arisen.   Here is a dilemma, either horn of which bears a threatening aspect to the exclusive supporter of "Natural

---

[1] See "Ann. and Mag. of Nat. Hist." for August 1870, p. 140.

[2] See " Proceedings of the Royal Institution," vol. v. part iv. p. 278 : Report of a Lecture delivered February 7, 1868.   Also "Quarterly Journal of the Geological Society," February 1870 :   "Contributions to the Anatomy and Taxonomy of the Dinosauria."

Selection," and between which it seems somewhat difficult to choose.

It has been suggested to me that this difficulty may be evaded by considering pterodactyles and carinate birds as independent branches from one side of an ancient common trunk, while similarly the Dinosauria and struthious birds are taken to be independent branches from the other side of the same common trunk ; the two kinds of birds resembling each other so much on account of their later development from that trunk as compared with the development of the reptilian forms. But to this it may be replied that the ancient common stock could not have had at one and the same time a shoulder structure of *both kinds.* It must have been that of the struthious birds or that of the carinate birds, or something different from both. If it was that of the struthious birds, how did the pterodactyles and carinate birds independently arrive at the very same divergent structure ? If it was that of the carinate birds, how did the struthious birds and Dinosauria independently agree to differ ? Finally, if it was something different from either, how did the carinate birds and pterodactyles take on independently one special common structure when disagreeing in so many; while the struthious birds, agreeing in many points with the Dinosauria, agree yet more with the carinate birds ? Indeed by no arrangement of branches from a stem can the difficulty be evaded.

Professor Huxley seems inclined [1] to cut the Gordian knot by considering the shoulder structure of the pterodactyle as independently educed, and having relation to physiology only. This conception is one which harmonizes completely with the views here advocated, and with those of Mr. Herbert Spencer, who also calls in direct modification to the aid of "Natural Selection." That merely minute, indefinite variations in all directions should unaided have independently built up the

[1] "Proceedings of Geological Society," November 1869, p. 38.

shoulder structure of the pterodactyles and carinate birds, and have laterally depressed their optic lobes, at a time so far back as the deposition of the Oolite strata,[1] is a coincidence of the highest improbability ; but that an innate power and evolutionary law, aided by the corrective action of " Natural Selection," should have furnished like needs with like aids, is not at all improbable. The difficulty does not tell against the theory of evolution, but only against the specially Darwinian form of it. Now this form has never been expressly adopted by Professor

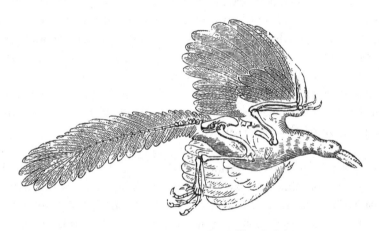

THE ARCHEOPTERYX (of the Oolite strata).

Huxley ; so far from it, in his lecture on this subject at the Royal Institution before referred to, he observes,[2] " I can testify, from personal experience, it is possible to have a complete faith in the general doctrine of evolution, and yet to hesitate in accepting the Nebular, or the Uniformitarian, or the Darwinian hypotheses in all their integrity and fulness."

It is quite consistent, then, in the Professor to explain the

---

[1] The archeopteryx of the oolite has the true carinate shoulder structure.
[2] " Proceedings of the Royal Institution," vol. v. p. 279.

difficulty as he does ; but it would not be similarly so with an absolute and pure Darwinian.

Yet stronger arguments of an analogous kind are, however, to be derived from the highest organs of sense. In the most perfectly organized animals—those namely which, like ourselves, possess a spinal column—the internal organs of hearing consist of two more or less complex membranous sacs (containing calcareous particles—otoliths), which are primitively or permanently lodged in two chambers, one on each side of the cartilaginous skull. The primitive cartilaginous cranium supports and protects the base of the brain, and the auditory nerves pass from that brain into the cartilaginous chambers to reach the auditory sacs. These complex arrangements of parts could not have been evolved by " Natural Selection," *i.e.* by minute accidental variations, except by the action of such through a vast period of time ; nevertheless, it was fully evolved at the time of the deposition of the upper Silurian rocks.

Cuttle-fishes (*Cephalopoda*) are animals belonging to the molluscous primary division of the animal kingdom, which division contains animals formed upon a type of structure utterly remote from that on which the animals of the higher division provided with a spinal column are constructed. And indeed no transitional form (tending even to bridge over the chasm between these two groups) has ever yet been discovered, either living or in a fossilized condition.[1]

Nevertheless, in the two-gilled Cephalopods (*Dibranchiata*) we find the brain supported and protected by a cartilaginous cranium. In the base of this cranium are two cartilaginous chambers. In each chamber is a membranous sac containing an otolith, and the auditory nerves pass from the cerebral ganglia into the cartilaginous chambers to reach the auditory sacs. Moreover,

[1] This remark is made without prejudice to possible affinities in the direction of the Ascidians,—an affinity which, if real, would be irrelevant to the question here discussed.

it has been suggested by Professor Owen that sinuosities between processes projecting from the inner wall of each chamber "seem to be the first rudiments of those which, in the higher classes (*i.e.* in animals with a spinal column), are extended in the form of canals and spiral chambers, within the substance of the dense nidus of the labyrinth."[1]

A

B

CUTTLE-FISH.

A. Ventral aspect.   B. Dorsal aspect.

Here, then, we have a wonderful coincidence indeed; two highly complex auditory organs, marvellously similar in structure, but which must nevertheless have been developed in entire and complete independence one of the other! It would be difficult to calculate the odds against the independent occurrence and conservation of two such complex series of merely accidental and minute haphazard variations. And it can never

[1] "Lectures on the Comp. Anat. of the Invertebrate Animals," 2nd edit. 1855, p. 619; and Todd's "Cyclopædia of Anatomy." vol. i. p. 554.

be maintained that the sense of hearing could not be efficiently subserved otherwise than by such sacs, in cranial cartilaginous capsules so situated in relation to the brain, &c.

Our wonder, moreover, may be increased when we recollect that the two-gilled cephalopods have not yet been found below the lias, where they at once abound ; whereas the four-gilled cephalopods are Silurian forms. Moreover, the absence is in this case significant in spite of the imperfection of the geological record, because when we consider how many individuals of various kinds of four-gilled cephalopods have been found, it is fair to infer that at the least a certain small percentage of dibranchs would also have left traces of their presence had they existed. Thus it is probable that some four-gilled form was the progenitor of the dibranch cephalopods. Now the four-gilled kinds (judging from the only existing form, the nautilus) had the auditory organ in a very inferior condition of development to what we find in the dibranch ; thus we have not only evidence of the independent high development of the organ in the former, but also evidence pointing towards a certain degree of comparative rapidity in its development.

Such being the case with regard to the organ of hearing, we have another yet stronger argument with regard to the organ of sight, as has been well pointed out by Mr. J. J. Murphy.[1] He calls attention to the fact that the eye must have been perfected in at least " three distinct lines of descent," alluding not only to the molluscous division of the animal kingdom, and the division provided with a spinal column, but also to a third primary division, namely, that which includes all insects, spiders, crabs, &c., which are spoken of as Annulosa, and the type of whose structure is as distinct from that of the molluscous type on the one hand, as it is from that of the type with a spinal column (*i.e.* the vertebrate type) on the other.

[1] See "Habit and Intelligence," vol. i. p. 321.

In the cuttle-fishes we find an eye even more completely constructed on the vertebrate type than is the ear. Sclerotic, retina, choroid, vitreous humour, lens, aqueous humour, all are present. The correspondence is wonderfully complete, and there can hardly be any hesitation in saying that for such an exact, prolonged, and correlated series of similar structures to have been brought about in two independent instances by merely indefinite and minute accidental variations, is an improbability which amounts practically to impossibility. Moreover, we have here again the same imperfection of the four-gilled cephalopod, as compared with the two-gilled, and therefore (if the latter proceeded from the former) a similar indication of a certain comparative rapidity of development. Finally, and this is perhaps one of the most curious circumstances, the process of formation appears to have been, at least in some respects, the same in the eyes of these molluscous animals as in the eyes of vertebrates. For in these latter the cornea is at first perforated, while different degrees of perforation of the same part are presented by different adult cuttle-fishes—large in the calamaries, smaller in the octopods, and reduced to a minute foramen in the true cuttle-fish sepia.

Some may be disposed to object that the conditions requisite for effecting vision are so rigid that similar results in all cases must be independently arrived at. But to this objection it may well be replied that Nature herself has demonstrated that there is no such necessity as to the details of the process. For in the higher Annulosa, such as the dragon-fly, we meet with an eye of an unquestionably very high degree of efficiency, but formed on a type of structure only remotely comparable with that of the fish or the cephalopod. The last-named animal might have had an eye as efficient as that of a vertebrate, but formed on a distinct type, instead of being another edition, as it were, of the very same structure.

In the beginning of this chapter examples have been given of the very diverse mode in which similar results have in many instances been arrived at; on the other hand, we have in the fish and the cephalopod not only the eye, but at one and the same time the ear also similarly evolved, yet with complete independence.

Thus it is here contended that the similar and complex structures of both the highest organs of sense, as developed in the vertebrates on the one hand, and in the mollusks on the other, present us with residuary phenomena for which "Natural Selection" alone is quite incompetent to account. And that these same phenomena must therefore be considered as conclusive evidence for the action of some other natural law or laws conditioning the simultaneous and independent evolution of these harmonious and concordant adaptations.

Provided with this evidence, it may be now profitable to enumerate other correspondences, which are not perhaps in themselves inexplicable by Natural Selection, but which are more readily to be explained by the action of the unknown law or laws referred to—which action, as its necessity has been demonstrated in one case, becomes *a priori* probable in the others.

Thus the great oceanic Mammalia—the whales—show striking resemblances to those prodigious, extinct, marine reptiles, the

SKELETON OF AN ICHTHYOSAURUS.

Ichthyosauria, and this not only in structures readily referable to similarity of habit, but in such matters as greatly elongated premaxillary bones, together with the concealment of certain bones of the skull by other cranial bones.

Again, the aërial mammals, the bats, resemble those flying reptiles of the secondary epoch, the pterodactyles; not only to a certain extent in the breast-bone and mode of supporting the flying membrane, but also in the proportions of different parts of the spinal column and the hinder (pelvic) limbs.

Also bivalve shell-fish (*i.e.* creatures of the mussel, cockle, and oyster class, which receive their name from the body being protected by a double shell, one valve of which is placed on each side) have their two shells united by one or two powerful muscles, which pass directly across from one shell to the other, and which are termed "adductor muscles" because by their contraction they bring together the valves and so close the shell.

Now there are certain animals which belong to the crab and lobster class (Crustacea)—a class constructed on an utterly different

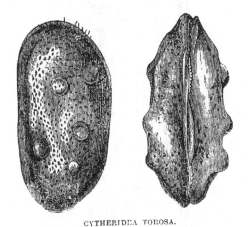

CYTHERIDEA TOROSA.

[An ostracod (Crustacean), externally like a bivalve shell-fish (Lamellibranch).

type from that on which the bivalve shell-fish are constructed—which present a very curious approximation to both the form and, in a certain respect, the structure of true bivalves. Allusion is here made to certain small Crustacea—certain phyllopods and ostracods—which have the hard outer coat of their thorax so modified as to look wonderfully like a bivalve shell, although its nature and composition are quite different. But this is by no means all,—not only is there this external resemblance between

A POLYZOON WITH BIRD'S-HEAD PROCESSES.

the thoracic armour of the crustacean and the bivalve shell, but the two sides of the ostracod and phyllopod thorax are connected together also by an adductor muscle !

The pedicellariæ of the echinus have been already spoken. of, and the difficulty as to their origin from minute, fortuitous, indefinite variations has been stated. But structures essentially similar (called avicularia, or "bird's-head processes") are developed from the surface of the compound masses of certain of the highest of the polyp-like animals (viz. the Polyzoa or, as they are sometimes called, the Bryozoa).

These compound animals have scattered over the surface of their bodies minute processes, each of which is like the head of a bird, with an upper and lower beak, the whole supported on a slender neck. The beak opens and shuts at intervals, like the jaws

BIRD'S-HEAD PROCESSES VERY GREATLY ENLARGED.

of the pedicellariæ of the echinus, and there is altogether, in general principle, a remarkable similarity between the structures. Yet the echinus can have, at the best, none but the most distant genetic relationship with the Polyzoa. We have here again therefore complex and similar organs of diverse and independent origin.

In the highest class of animals (the Mammalia) we have almost always a placental mode of reproduction, *i.e.* the blood of the fœtus is placed in nutritive relation with the blood of the mother by means of vascular prominences. No trace of such a structure exists in any bird or in any reptile, and yet it crops out again in certain sharks. There indeed it might well be supposed to end, but, marvellous as it seems, it reappears in very lowly creatures; namely, in certain of the ascidians, sometimes called tunicaries or sea-squirts.

Now, if we were to concede that the ascidians were the common ancestors[1] of both these sharks and of the higher mammals, we should be little, if any, nearer to an explanation of the phenomenon by means of "Natural Selection," for in the sharks in question the vascular prominences are developed from one foetal structure (the umbilical vesicle), while in the higher mammals they are developed from quite another part, viz. the allantois.

Upper Figure—Antechinus minutissimus (*implacental*).
Lower Figure—Mus delicatulus (*placental*).

So great, however, is the number of similar, but apparently independent, structures, that we suffer from a perfect *embarras de richesses*. Thus, for example, we have the convoluted windpipe

[1] A view recently propounded by Kowalewsky.

of the sloth, reminding us of the condition of the windpipe in birds; and in another mammal, allied to the sloth, namely the great ant-eater (Myrmecophaga), we have again an ornithic character in its horny gizzard-like stomach.   In man and the highest apes the cæcum has a vermiform appendix, as it has also in the wombat !

Also the similar forms presented by the crowns of the teeth in some seals, in certain sharks, and in some extinct Cetacea may be referred to ; as also the similarity of the beak in birds, some reptiles, in the tadpole, and cuttle-fishes.   As to entire external form, may be adduced the wonderful similarity between a true mouse (*Mus delicatulus*) and a small marsupial, pointed out by Mr. Andrew Murray in his work on the "Geographical Distribution of Mammals," p. 53, and represented in the frontispiece by figures copied from Gould's "Mammals of Australia ;" but instances enough for the present purpose have been already quoted.

Additional reasons for believing that similarity of structure is produced by other causes than merely by "Natural Selection" are furnished by certain facts of zoological geography, and by a similarity in the mode of variation being sometimes extended to several species of a genus, or even to widely different groups ; while the restriction and the limitation of such similarity are often not less remarkable.   Thus Mr. Wallace says,[1] as to local influence : " Larger or smaller districts, or even single islands, give a special character to the majority of their Papilionidæ. For instance :—1. The species of the Indian region (Sumatra, Java, and Borneo) are almost invariably smaller than the allied species inhabiting Celebes and the Moluccas. 2. The species of New Guinea and Australia are also, though in a less degree, smaller than the nearest species or varieties of the Moluccas. 3. In the Moluccas themselves the species of Amboyna are the largest. 4. The species of Celebes equal or even surpass in size those

---

[1] "Natural Selection," p. 167.

of Amboyna. 5. The species and varieties of Celebes possess a striking character in the form of the anterior wings, different from that of the allied species and varieties of all the surrounding islands. 6. Tailed species in India or the Indian region become tailless as they spread eastward through the Archipelago. 7. In Amboyna and Ceram the females of several species are dull-coloured, while in the adjacent islands they are more brilliant." Again:[1] "In Amboyna and Ceram the female of the large and handsome *Ornithoptera Helena* has the large patch on the hind wings constantly of a pale dull ochre or buff colour; while in the scarcely distinguishable varieties from the adjacent islands, of Bouru and New Guinea, it is of a golden yellow, hardly inferior in brilliancy to its colour in the male sex. The female of *Ornithoptera Priamus* (inhabiting Amboyna and Ceram exclusively) is of a pale dusky brown tint, while in all the allied species the same sex is nearly black, with contracted white markings. As a third example, the female of *Papilio Ulysses* has the blue colour obscured by dull and dusky tints, while in the closely allied species from the surrounding islands, the females are of almost as brilliant an azure blue as the males. A parallel case to this is the occurrence, in the small islands of Goram, Matabello, Ké, and Aru, of several distinct species of Euploea and Diadema, having broad bands or patches of white, which do not exist in any of the allied species from the larger islands. These facts seem to indicate some local influence in modifying colour, as unintelligible and almost as remarkable as that which has resulted in the modifications of form previously described."

After endeavouring to explain some of the facts in a way to be noticed directly, Mr. Wallace adds:[2] "But even the conjectural explanation now given fails us in the other cases of local modification. Why the species of the Western Islands should be smaller than those further east; why those of Amboyna

---

[1] " Natural Selection," p. 173.   [2] Ibid. p. 177.

should exceed in size those of Gilolo and New Guinea; why
the tailed species of India should begin to lose that appendage
in the islands, and retain no trace of it on the borders of
the Pacific; and why, in three separate cases, the females of
Amboyna species should be less gaily attired than the corresponding females of the surrounding islands, are questions
which we cannot at present attempt to answer. That they
depend, however, on some general principle is certain, because
analogous facts have been observed in other parts of the world.
Mr. Bates informs me that, in three distinct groups, Papilios,
which, on the Upper Amazon, and in most other parts of South
America, have spotless upper wings, obtain pale or white spots
at Pará and on the Lower Amazon, and also that the Æneas
group of Papilios never have tails in the equatorial regions and
the Amazon valley, but gradually acquire tails in many cases as
they range towards the northern or southern tropic. Even in
Europe we have somewhat similar facts, for the species and
varieties of butterflies peculiar to the Island of Sardinia are
generally smaller and more deeply coloured than those of the
mainland, and the same has been recently shown to be the case
with the common tortoiseshell butterfly in the Isle of Man;
while *Papilio Hospiton*, peculiar to the former island, has lost
the tail, which is a prominent feature of the closely allied
P. Machaon.

. " Facts of a similar nature to those now brought forward would
no doubt be found to occur in other groups of insects, were
local faunas carefully studied in relation to those of the surrounding countries; and they seem to indicate that climate and
other physical causes have, in some cases, a very powerful effect
in modifying specific form and colour, and thus directly aid in
producing the endless variety of nature."

With regard to butterflies of Celebes belonging to different
families, they present " a peculiarity of outline which distinguishes them at a glance from those of any other part of the

world :"[1] it is that the upper wings are generally more elongated
and the anterior margin more curved. Moreover, there is, in

OUTLINES OF WINGS OF BUTTERFLIES OF CELEBES COMPARED WITH THOSE OF
ALLIED SPECIES ELSEWHERE.

Outer outline, *Papilio gigon*, of Celebes. Inner outline, *P. demolion*, of Singapore
and Java.—2. Outer outline, *P. miletus*, of Celebes. Inner outline, *P. sarpedon*,
India.—3. Outer outline, *Tachyris zarinda*, Celebes. Inner outline, *T. nero*.

most instances, near the base an abrupt bend or elbow,
which in some species is very conspicuous. Mr. Wallace

[1] "Malay Archipelago," vol. i. p. 439.

endeavours to explain this phenomenon by the supposed presence at some time of special persecutors of the modified forms, supporting the opinion by the remark that small, obscure, very rapidly flying and mimicked kinds have not had the wing modified. Such an enemy occasioning increased powers of flight, or rapidity in turning, he adds, "one would naturally suppose to be an insectivorous bird; but it is a remarkable fact that most of the genera of fly-catchers of Borneo and Java on the one side, and of the Moluccas on the other, are almost entirely absent from Celebes. Their place seems to be supplied by the caterpillar-catchers, of which six or seven species are known from Celebes, and are very numerous in individuals. We have no positive evidence that these birds pursue butterflies on the wing, but it is highly probable that they do so when other food is scarce. Mr. Bates suggested to me that the larger dragon-flies prey upon butterflies, but I did not notice that they were more abundant in Celebes than elsewhere."[1]

Now, every opinion or conjecture of Mr. Wallace is worthy of respectful and attentive consideration, but the explanation suggested and before referred to hardly seems a satisfactory one. What the past fauna of Celebes may have been is as yet conjectural. Mr. Wallace tells us that now there is a remarkable *scarcity* of fly-catchers, and that their place is supplied by birds of which it can only be said that it is "highly probable" that they chase butterflies "when other food is scarce." The quick eye of Mr. Wallace failed to detect them in the act, as also to note any unusual abundance of other insectivorous forms, which therefore, considering Mr. Wallace's zeal and powers of observation, we may conclude do not exist. Moreover, even if there ever has been an abundance of such, it is by no means certain that they would have succeeded in producing the conformation in question, for the effect of this peculiar curvature on flight is by no means clear. We have here, then, a structure

[1] "Natural Selection," p. 177.

hypothetically explained by an uncertain property induced by a cause the presence of which is only conjectural.

Surely it is not unreasonable to class this instance with the others before given, in which a common modification of form or colour coexists with a certain geographical distribution quite independently of the destructive agencies of animals. If physical causes connected with locality can abbreviate or annihilate the tails of certain butterflies, why may not similar causes produce an elbow-like prominence on the wings of other butterflies? There are many such instances of simultaneous modification. Mr. Darwin himself [1] quotes Mr. Gould as believing that birds of the same species are more brightly coloured under a clear atmosphere, than when living on islands or near the coast. Mr. Darwin also informs us that Wollaston is convinced that residence near the sea affects the colour of insects; and finally, that Moquin-Tandon gives a list of plants which, when growing near the sea-shore, have their leaves in some degree fleshy, though not so elsewhere. In his work on "Animals and Plants under Domestication," [2] Mr. Darwin refers to M. Costa as having (in *Bull. de la Soc. Imp. d'Acclimat.* tome viii. p. 351) stated " that young shells taken from the shores of England and placed in the Mediterranean at once altered their manner of growth, and formed prominent diverging rays *like those on the shells of the proper Mediterranean oyster ;*" also to Mr. Meehan, as stating (*Proc. Acad. Nat. Sc. of Philadelphia,* Jan. 28, 1862) " that twenty-nine kinds of American trees all differ from their nearest European allies in *a similar manner,* leaves less toothed, buds and seeds smaller, fewer branchlets," &c. These are striking examples indeed !

But cases of simultaneous and similar modifications abound on all sides. Even as regards our own species there is a very generally admitted opinion that a new type has been developed in the United States, and this in about a couple of

---

[1] " Origin of Species," 5th edition, p. 166.   [2] Vol. ii. p. 280.

centuries only, and in a vast multitude of individuals of diverse ancestry. The instances here given, however, must suffice, though more could easily be added.

It may be well now to turn to groups presenting similar variations, not through, but independently of, geographical distribution, and, as far as we know, independently of conditions other than some peculiar nature and tendency (as yet unexplained) common to members of such groups, which nature and tendency seem to induce them to vary in certain definite lines or directions which are different in different groups.

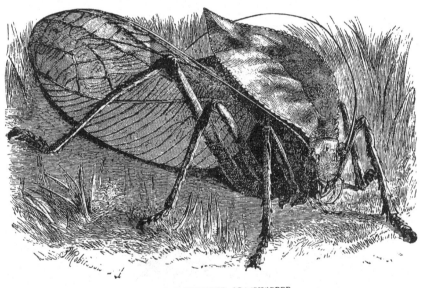

THE GREAT SHIELDED GRASSHOPPER.

Thus with regard to the group of insects, of which the walking leaf is a member, Mr. Wallace observes :[1] "The *whole family*[2] of the Phasmidæ, or spectres, to which this insect belongs, is more or less imitative, and a great number of the species are called

[1] See " Natural Selection," p. 64.
[2] The italics are not Mr. Wallace's.

'walking-stick insects,' from their singular resemblance to twigs and branches."

Again, Mr. Wallace[1] tells us of no less than four kinds of orioles, which birds mimic, more or less, four species of a genus of honey-suckers, the weak orioles finding their profit in being mistaken by certain birds of prey for the strong, active, and gregarious honey-suckers.   Now, many other birds would be benefited by similar mimicry, which is none the less confined, in this part of the world, to the oriole genus.   It is true that the absence of mimicry in other forms may be explained by their

THE SIX-SHAFTED BIRD OF PARADISE.

possessing some other (as yet unobserved) means of preservation. But it is nevertheless remarkable, not so much that one species should mimic, as that no less than four should do so in different

[1] "Malay Archipelago," vol. ii. p. 150 ; and "Natural Selection," p. 104.

ways and degrees, all these four belonging to *one and the same genus*.

In other cases, however, there is not even the help of protective action to account for the phenomenon. Thus we have the wonderful birds of Paradise,[1] which agree in developing plumage

THE LONG-TAILED BIRD OF PARADISE.

unequalled in beauty, but a beauty which, as to details, is of different kinds, and produced in different ways in different species. To develop "beauty and singularity of plumage" is a character of the group, but not of any one definite kind, to be explained merely by inheritance.

[1] See "Malay Archipelago," vol. ii. chap. xxxviii.

THE RED BIRD OF PARADISE.

Again, we have the very curious horned flies,[1] which agree indeed in a common peculiarity, but in one singularly different in detail, in different species and not known to have any protecting effect.

Amongst plants, also, we meet with the same peculiarity. The great group of Orchids presents a number of species

[1] Loc. cit. p. 314.

which offer strange and bizarre approximations to different

HORNED FLIES.

animal forms, and which have often the appearance of cases of mimicry, as it were in an incipient stage.

THE MAGNIFICENT BIRD OF PARADISE.

The number of similar instances which could be brought forward from amongst animals and plants is very great, but the examples given are, it is hoped, amply sufficient to point towards the conclusion which other facts will, it is thought, establish, viz. that there are causes operating (in the evocation of these harmonious diverging resemblances) other than "Natural Selection," or heredity, and other even than merely geographical, climatal, or any simply external conditions.

Many cases have been adduced of striking likenesses between different animals, not due to inheritance; but this should be the less surprising, in that the very same individual presents us with likenesses between different parts of its body (*e.g.*, between the several joints of the backbone), which are certainly not so explicable. This, however, leads to a rather large subject, which will be spoken of in the eighth chapter of the present work. Here it will be enough to affirm (leaving the proof of the assertion till later) that parts are often homologous which have no direct genetic relationship,—a fact which harmonizes well with the other facts here given, but which "Natural Selection," pure and simple, seems unable to explain.

But surely the independent appearance of similar organic forms is what we might expect, *a priori*, from the independent appearance of similar inorganic ones. As Mr. G. H. Lewes well observes,[1] "We do not suppose the carbonates and phosphates found in various parts of the globe—we do not suppose that the families of alkaloids and salts have any nearer kinship than that which consists in the similarity of their elements, and the conditions of their combination. Hence, in organisms, as in salts, morphological identity may be due to a community of causal connexion, rather than community of descent.

"Mr. Darwin justly holds it to be incredible that individuals identically the same should have been produced through Natural Selection from parents *specifically distinct,* but he will not deny that identical forms may issue from parents *genetically distinct,*

---

[1] *Fortnightly Review*, New Series, vol. iii (April 1868), p. 372.

when these parent forms and the conditions of production are identical.   To deny this would be to deny the law of causation."

Professor Huxley has, however, suggested[1] that such mineral identity may be explained by applying also to minerals a law of descent; that is, by considering such similar forms as the descendants of atoms which inhabited one special part of the primitive nebular cosmos, each considerable space of which may be supposed to have been under the influence of somewhat different conditions.

Surely, however, there can be no real parity between the relationship of existing minerals to nebular atoms, and the relationship of existing animals and plants to the earliest organisms.   In the first place, the latter have produced others by generative multiplication, which mineral atoms never did.   In the second, existing animals and plants spring from the living tissues of preceding animals and plants, while existing minerals spring from the chemical affinity of separate elements.   Carbonate of soda is not formed, by a process of reproduction, from other carbonate of soda, but directly by the suitable juxtaposition of carbon, oxygen, and sodium.

Instead of approximating animals and minerals in the mode suggested, it may be that they are to be approximated in quite a contrary fashion; namely, by attributing to mineral species an internal innate power.   For, as we must attribute to each elementary atom an innate power and tendency to form (under the requisite external conditions) certain unions with other atoms, so we may attribute to certain mineral species — as crystals—an innate power and tendency to exhibit (the proper conditions being supplied) a definite and symmetrical external form.   The distinction between animals and vegetables on the one hand, and minerals on the other, is that, while in the organic world close similarity is the result sometimes of inheritance, sometimes of direct production independently of parental

[1] " Lay Sermons," p. 339.

action, in the inorganic world the latter is the constant and
only mode in which such similarity is produced.

When we come to consider the relations of species to space—
in other words, the geographical distribution of organisms—it
will be necessary to return somewhat to the subject of the independent origin of closely similar forms, in regard to which
some additional remarks will be found towards the end of the
seventh chapter.

In this third chapter an effort has been made to show that
while on the Darwinian theory concordant variations are
extremely improbable, yet Nature presents us with abundant
examples of such; the most striking of which are, perhaps,
the higher organs of sense. Also that an important influence
is exercised by conditions connected with geographical distribution, but that a deeper-seated influence is at work, which is
hinted at by those special tendencies in definite directions, which
are the properties of certain groups. Finally, that these facts,
when taken together, afford strong evidence that " Natural
Selection" has not been the exclusive or predominant cause of
the various organic structural peculiarities. This conclusion has
also been re-enforced by the consideration of phenomena presented to us by the inorganic world.

CHAPTER IV.

MINUTE AND GRADUAL MODIFICATIONS.

There are difficulties as to minute modifications, even if not fortuitous.—
Examples of sudden and considerable modifications of different kinds.—
Professor Owen's view.—Mr. Wallace.—Professor Huxley.—Objections
to sudden changes.—Labyrinthodont.—Potto.—Cetacea.—As to origin
of bird's wing.—Tendrils of climbing plants.—Animals once supposed
to be connecting links.—Early specialization of structure.—Macrauchenia.—Glyptodon.—Sabre-toothed tiger.—Conclusion.

Not only are there good reasons against the acceptance of
the exclusive operation of "Natural Selection" as the one
means of specific origination, but there are difficulties in the
way of accounting for such origination by the sole action of
modifications which are infinitesimal and minute, whether
fortuitous or not.

Arguments may yet be advanced in favour of the view that new
species have from time to time manifested themselves with suddenness, and by modifications appearing at once (as great in
degree as are those which separate *Hipparion* from *Equus*), the
species remaining stable in the intervals of such modifications:
by stable being meant that their variations only extend for a
certain degree in various directions, like oscillations in a
stable equilibrium. This is the conception of Mr. Galton,[1]
who compares the development of species with a many facetted

" Hereditary Genius, an Inquiry into its Laws," &c.   By Francis Galton,
F. R. S.   (London : Macmillan.)

H

spheroid tumbling over from one facet, or stable equilibrium, to another. The existence of internal conditions in animals corresponding with such facets is denied by pure Darwinians, but it is contended in this work, though not in this chapter, that something may also be said for their existence.

The considerations brought forward in the last two chapters, namely, the difficulties with regard to incipient and closely similar structures respectively, together with palæontological considerations to be noticed later, appear to point strongly in the direction of sudden and considerable changes. This is notably the case as regards the young oysters already mentioned, which were taken from the shores of England and placed in the Mediterranean, and at once altered their mode of growth and formed prominent diverging rays, *like those of the proper Mediterranean oyster ;* as also the twenty-nine kinds of American trees, all differing from their nearest European allies *similarly*—"leaves less toothed, buds and seeds smaller, fewer branchlets," &c. To these may be added other facts given by Mr. Darwin. Thus he says, "that climate, to a certain extent, directly modifies the form of dogs."[1]

The Rev. R. Everett found that setters at Delhi, though most carefully paired, yet had young with "nostrils more contracted, noses more pointed, size inferior, and limbs more slender." Again, cats at Mombas, on the coast of Africa, have short stiff hairs instead of fur, and a cat at Algoa Bay, when left only eight weeks at Mombas, "underwent a complete metamorphosis, having parted with its sandy-coloured fur."[2] The conditions of life seem to produce a considerable effect on horses, and instances are given by Mr. Darwin of pony breeds[3] having independently arisen in different parts of the world, possessing a certain similarity in their physical condi-

---

[1] "Animals and Plants under Domestication," vol. i. p. 37.
[2] Ibid. p. 47.   [3] Ibid. p. 52.

tions. Also changes due to climate may be brought about at once in a second generation, though no appreciable modification is shown by the first. Thus " Sir Charles Lyell mentions that some Englishmen, engaged in conducting the operations of the Real del Monte Company in Mexico, carried out with them some greyhounds of the best breed to hunt the hares which abound in that country. It was found that the greyhounds could not support the fatigues of a long chase in this attenuated atmosphere, and before they could come up with their prey they lay down gasping for breath ; but these same animals have produced whelps, which have grown up, and are not in the least degree incommoded by the want of density in the air, but run down the hares with as much ease as do the fleetest of their race in this country." [1]

We have here no action of " Natural Selection ;" it was not that certain puppies happened accidentally to be capable of enduring more rarefied air, and so survived, but the offspring were directly modified by the action of surrounding conditions. Neither was the change elaborated by minute modifications in many successive generations, but appeared at once in the second.

With regard once more to sudden alterations of form, Nathusius is said to state positively as to pigs,[2] that the result of common experience and of his experiments was that rich and abundant food, given during youth, tends by some direct action to make the head broader and shorter. Curious jaw appendages often characterize Normandy pigs, according to M. Eudes Deslongchamps. Richardson figures these appendages on the old " Irish greyhound pig," and they are said by Nathusius to appear occasionally in all the long-eared races. Mr. Darwin observes,[3] " As no wild pigs are known to have analogous

---

[1] Carpenter's " Comparative Physiology," p. 987, quoted by Mr. J. J Murphy, " Habit and Intelligence," vol. i p. 171.

" Animals and Plants under Domestication,' vol. i. p. 72.

[3] Ibid. p. 76.

appendages, we have at present no reason to suppose that their appearance is due to reversion; and if this be so, we are forced to admit that somewhat complex, though apparently useless structures may be suddenly developed without the aid of selection." Again, "Climate directly affects the thickness of the skin and hair" of cattle.[1] In the English climate an individual Porto Santo rabbit[2] recovered the proper colour of its fur in rather less than four years. The effect of the climate of India on the turkey is considerable. Mr. Blyth[3] describes it as being much degenerated in size, "utterly incapable of rising on the wing," of a black colour, and "with long pendulous appendages over the beak enormously developed." Mr. Darwin again tells us that there has suddenly appeared in a bed of common broccoli a peculiar variety, faithfully transmitting its newly acquired and remarkable characters;[4] also that there have been a rapid transformation and transplantation of American varieties of maize with a European variety;[5] that certainly "the Ancon and Manchamp breeds of sheep," and that (all but certainly) Niata cattle, turnspit and pug dogs, jumper and frizzled fowls, short-faced tumbler pigeons, hook-billed ducks, &c., and a multitude of vegetable varieties, have suddenly appeared in nearly the same state as we now see them.[6] Lastly, Mr. Darwin tells us, that there has been an occasional development (in five distinct cases) in England of the "japanned" or "black-shouldered peacock" (*Pavo nigripennis*), a distinct species, according to Dr. Sclater,[7] yet arising in Sir J. Trevelyan's flock composed entirely of the common kind, and increasing, "*to the extinction of the previously existing breed.*"[8] Mr. Darwin's only explanation of the phenomena (on the supposition of the

---

[1] "Animals and Plants under Domestication," vol. i. p. 71.
[2] Ibid. p. 114.   [3] Quoted, Ibid. p. 274.   [4] Ibid. p. 324.
[5] Ibid. p. 322.   [6] Ibid. vol. ii. p. 414.
[7] Proc. Zool. Soc. of London, April 24, 1860.
[8] "Animals and Plants under Domestication," vol. i. p. 291.

species being distinct) is by reversion, owing to a supposed
ancestral cross. But he candidly admits, " I have heard of
no other such case in the animal or vegetable kingdom." On
the supposition of its being only a variety, he observes, " The
case is the most remarkable ever recorded of the abrupt
appearance of a new form, which so closely resembles a true
species, that it has deceived one of the most experienced of
living ornithologists."

As to plants, M. C. Naudin[1] has given the following instances
of the sudden origination of apparently permanent forms. " The
first case mentioned is that of a poppy, which took on a remarkable variation in its fruit—a crown of secondary capsules being
added to the normal central capsule. A field of such poppies
was grown, and M. Göppert, with seed from this field, obtained
still this monstrous form in great quantity. Deformities of
ferns are sometimes sought after by fern-growers. They are
now always obtained by taking spores from the abnormal
parts of the monstrous fern; from which spores ferns presenting the same peculiarities invariably grow. . . . . The most
remarkable case is that observed by Dr. Godron, of Nancy.
In 1861 that botanist observed, amongst a sowing of
Datura tatula, the fruits of which are very spinous, a single
individual of which the capsule was perfectly smooth. The
seeds taken from this plant all furnished plants having the
character of this individual. The fifth and sixth generations are now growing without exhibiting the least tendency
to revert to the spinous form. More remarkable still, when
crossed with the normal *Datura tatula*, hybrids were produced,
which, in the second generation, reverted to the original types,
as true hybrids do."

There are, then, abundant instances to prove that considerable

---

[1] Extracted by J. J. Murphy, vol. i. p. 197, from the *Quarterly
Journal of Science*, of October 1867, p. 527.

modifications may suddenly develop themselves, either due to external conditions or to obscure internal causes in the organisms which exhibit them. Moreover, these modifications, from whatever cause arising, are capable of reproduction—the modified individuals "breeding true."

The question is whether new species have been developed by non-fortuitous variations which are insignificant and minute, or whether such variations have been comparatively sudden, and of appreciable size and importance? Either hypothesis will suit the views here maintained equally well (those views being opposed only to fortuitous, indefinite variations), but the latter is the more remote from the Darwinian conception, and yet has much to be said in its favour.

Professor Owen considers, with regard to specific origination, that natural history " teaches that the change would be sudden and considerable : it opposes the idea that species are transmitted by minute and slow degrees."[1] " An innate tendency to deviate from parental type, operating through periods of adequate duration," being " the most probable nature, cr way of operation of the secondary law, whereby species have been derived one from the other."[2]

Now, considering the number of instances adduced of sudden modifications in domestic animals, it is somewhat startling to meet with Mr. Darwin's dogmatic assertion that it is " a *false belief*" that natural species have often originated in the same abrupt manner. The belief *may* be false, but it is difficult to see how its falsehood can be positively asserted.

It is demonstrated by Mr. Darwin's careful weighings and measurements, that, though little used parts in domestic animals get reduced in weight and somewhat in size, yet that they show no inclination to become truly " rudimentary structures."

[1] "Anatomy of Vertebrates," vol. iii. p. 795.
[2] Ibid. p. 807.

Accordingly he asserts[1] that such rudimentary parts are formed "suddenly, by arrest of development" in domesticated animals, but in wild animals slowly. The latter assertion, however, is a *mere assertion ;* necessary, perhaps, for the theory of "Natural Selection," but as yet unproved by facts.

But why should not these changes take place suddenly in a state of nature ? As Mr. Murphy says,[2] " It may be true that we have no evidence of the origin of wild species in this way. But this is not a case in which negative evidence proves anything. We have never witnessed the origin of a wild species by any process whatever ; and if a species were to come suddenly into being in the wild state, as the Ancon Sheep did under domestication, how could you ascertain the fact ? If the first of a newly-begotten species were found, the fact of its discovery would tell nothing about its origin. Naturalists would register it as a very rare species, having been only once met with, but they would have no means of knowing whether it were the first or the last of its race."

To this Mr. Wallace has replied (in his review of Mr. Murphy's work in *Nature*[3]), by objecting that sudden changes could very rarely be useful, because each kind of animal is a nicely balanced and adjusted whole, any one sudden modification of which would in most cases be hurtful unless accompanied by other simultaneous and harmonious modifications. If, however, it is not unlikely that there is an innate tendency to deviate at certain times, and under certain conditions, it is no more unlikely that that innate tendency should be an harmonious one, calculated to simultaneously adjust the various parts of the organism to their new relations. The objection as to the sudden abortion of rudimentary organs may be similarly met.

Professor Huxley seems now disposed to accept the, at least

[1] " Animals and Plants under Domestication," vol. ii. p. 318.
[2] " Habit and Intelligence," vol. i. p. 344.
[3] See Dec. 2, 1869, vol. i. p. 132.

occasional, intervention of sudden and considerable variations. In his review of Professor Kölliker's[1] criticisms, he himself says,[2] "We greatly suspect that she" (*i.e.* Nature) "does make considerable jumps in the way of variation now and then,

MUCH ENLARGED HORIZONTAL SECTION OF THE TOOTH OF A LABYRINTHODON.

and that these saltations give rise to some of the gaps which appear to exist in the series of known forms."

In addition to the instances brought forward in the second chapter against the minute action of Natural Selection, may be

[1] "Über die Darwin'sche Schöpfungstheorie:" ein Vortrag, von Kölliker; Leipzig, 1864.   [2] See "Lay Sermons," p. 342.

mentioned such structures as the wonderfully folded teeth of the labyrinthodonts.   The marvellously complex structure of these organs is not merely unaccountable as due to Natural "Selection," but its production by insignificant increments of complexity is hardly less difficult to comprehend.

Similarly the aborted index of the Potto (*Perodicticus*) is a structure not likely to have been induced by minute changes; while, as to " Natural Selection," the reduction of the fore-finger to a mere rudiment is inexplicable indeed !   " How this muti-

HAND OF THE POTTO (PERODICTICUS), FROM LIFE.

lation can have aided in the struggle for life, we must confess, baffles our conjectures on the subject ; for that any very appreciable gain to the individual can have resulted from the slightly lessened degree of required nourishment thence resulting (*i.e.* from the suppression), seems to us to be an almost absurd proposition." [1]

Again, to anticipate somewhat, the great group of whales (Cetacea) was fully developed at the deposition of the Eocene strata.   On the other hand, we may pretty safely conclude that these animals were absent as late as the latest secondary rocks, so that their development could not have been so very slow, unless geological time is (although we shall presently see there are grounds to believe it is not) practically infinite.   It is quite true that it is, in general, very unsafe to infer the absence of any animal forms

[1] "Anatomy of the Lemuroidea." By James Murie, M.D., and St. George Mivart. Trans. Zool. Soc., March 1866, p. 91.

during a certain geological period, because no remains of them
have as yet been found in the strata then deposited : but in the
case of the Cetacea it is safe to do so; for, as Sir Charles Lyell
remarks,[1] they are animals, the remains of which are singularly
likely to have been preserved had they existed, in the same way
that the remains were preserved of the Ichthyosauri and Ple-

SKELETON OF A PLESIOSAURUS.

siosauri, which appear to have represented the Cetacea during
the secondary geological period.

As another example, let us take the origin of wings, such as
exist in birds. Here we find an arm, the bones of the hand of
which are atrophied and reduced in number, as compared with
those of most other Vertebrates. Now, if the wing arose from
a terrestrial or subaërial organ, this abortion of the bones could
hardly have been serviceable—hardly have preserved individuals
in the struggle for life. If it arose from an aquatic organ, like
the wing of the penguin, we have then a singular divergence
from the ordinary vertebrate fin-limb. In the ichthyosaurus, in
the plesiosaurus, in the whales, in the porpoises, in the seals, and
in others, we have shortening of the bones, but no reduction in
the number either of the fingers or of their joints, which are, on
the contrary, multiplied in Cetacea and the ichthyosaurus. And
even in the turtles we have eight carpal bones and five digits,

[1] " Principles of Geology," last edition, vol. i. p. 163.

while no finger has less than two phalanges. It is difficult, then, to believe that the Avian limb was developed in any other way than by a comparatively sudden modification of a marked and important kind.

How, once more, can we conceive the peculiar actions of the tendrils of some climbing plants to have been produced by minute modifications? These, according to Mr. Darwin,[1] oscillate till they touch an object, and then embrace it. It is stated by that observer, "that a thread weighing no more than the thirty-second of a grain, if placed on the tendril of the *Passiflora gracilis,* will cause it to bend; and merely to touch the tendril with a twig causes it to bend; but if the twig is at once

SKELETON OF AN ICHTHYOSAURUS.

removed, the tendril soon straightens itself. But the contact of other tendrils of the plant, or of the falling of drops of rain, do not produce these effects.[2] But some of the zoological and anatomical discoveries of late years tend rather to diminish than to augment the evidence in favour of minute and gradual modification. Thus all naturalists now admit that certain animals, which were at one time supposed to be connecting links between groups, belong altogether to one group, and not at all to the other. For example, the aye-aye[3] (*Chiromys Madagascariensis*)

[1] *Quarterly Journal of Science,* April 1866, pp. 257–8.
[2] "Habit and Intelligence," vol. i. p. 178.
[3] This animal belongs to the order Primates, which includes man, the apes, and the lemurs. The lemurs are the lower kinds of the order, and

was till lately considered to be allied to the squirrels, and was often classed with them in the rodent order, principally on account of its dentition; at the same time that its affinities to the lemurs and apes were admitted. The thorough investigation into its anatomy that has now been made, demonstrates that it has no more essential affinity to rodents than any other lemurine creature has.

THE AYE-AYE.

Bats were, by the earliest observers, naturally supposed to have a close relationship to birds, and cetaceans to fishes. It is almost superfluous to observe that all now agree that these mammals make not even an approach to either one or other of the two inferior classes.

differ much from the apes. They have their head-quarters in the Island of Madagascar. The aye-aye is a lemur, but it differs singularly from all its congeners, and still more from all apes. In its dentition it strongly approximates to the rodent (rat, squirrel, and guinea-pig) order, as it has two cutting teeth above, and two below, growing from permanent pulps, and in the adult condition has no canines.

In the same way it has been recently supposed that those extinct flying saurians, the pterodactyles, had an affinity with birds more marked than any other known animals. Now, however, as has been said earlier, it is contended that not only had they no such close affinity, but that other extinct reptiles had a far closer one.

The *amphibia* (*i.e.* frogs, toads, and efts) were long considered (and are so still by some) to be reptiles, showing an affinity to fishes. It now appears that they form with the latter one great group—the ichthyopsida of Professor Huxley—which differs widely from reptiles; while its two component classes (fishes and amphibians) are difficult to separate from each other in a thoroughly satisfactory manner.

If we admit the hypothesis of gradual and minute modification, the succession of organisms on this planet must have been a progress from the more general to the more special, and no doubt this has been the case in the majority of instances. Yet it cannot be denied that some of the most recently formed fossils show a structure singularly more generalized than any exhibited by older forms; while others are more specialized than are any allied creatures of the existing creation.

A notable example of the former circumstance is offered by macrauchenia—a hoofed animal, which was at first supposed to be a kind of great llama (whence its name)—the llama being a ruminant, which, like all the rest, has two toes to each foot. Now hoofed animals are divisible into two very distinct series, according as the number of functional toes on each hind foot is odd or even. And many other characters are found to go with this obvious one. Even the very earliest Ungulata show this distinction, which is completely developed and marked even in the Eocene palæotherium and anoplotherium found in Paris by Cuvier. The former of these has the toes odd (perissodactyle), the other has them even (artiodactyle).

Now, the macrauchenia, from the first relics of it which were found, was thought to belong, as has been said, to the even-toed division. Subsequent discoveries, however, seemed to give it an equal claim to rank amongst the perissodactyle forms. Others again inclined the balance of probability towards the artiodactyle. Finally, it appears that this very recently extinct beast presents a highly generalized type of structure, uniting in one organic form both artiodactyle and perissodactyle characters, and that in a manner not similarly found in any other known creature living, or fossil. At the same time the differentiation of artiodactyle and perissodactyle forms existed as long ago as in the period of the Eocene ungulata, and that in a marked degree, as has been before observed.

Again, no armadillo *now living* presents nearly so remarkable a speciality of structure as was possessed by the *extinct* glyptodon. In that singular animal the spinal column had most of its joints fused together, forming a rigid cylindrical rod, a modification, as far as yet known, absolutely peculiar to it.

In a similar way the *extinct* machairodus, or sabre-toothed

DENTITION OF THE SABRE-TOOTHED TIGER (MACHAIRODUS)

tiger, is characterized by a more highly differentiated and specially carnivorous dentition than is shown by any predacious beast of the *present day.* The specialization is of this kind. The grinding teeth (or molars) of beasts are divided into premolars and true molars. The premolars are molars which have deciduous vertical predecessors (or milk teeth), and any which are in front of such, *i.e.* between such and the canine tooth. The true molars are those placed behind the molars having deciduous vertical predecessors. Now, as a dentition becomes more distinctly carnivorous, so the hindmost molars and the foremost premolars disappear. In the existing cats this process is carried so far that in the upper jaw only one true molar is left on each side. In the machairodus there is no upper true molar at all, while the premolars are reduced to two, there being only these two teeth above, on each side, behind the canine.

Now, with regard to these instances of early specialization, as also with regard to the changed estimate of the degrees of affinity between forms, it is not pretended for a moment that such facts are irreconcilable with "Natural Selection." Nevertheless, they point in an opposite direction. Of course not only is it conceivable that certain antique types arrived at a high degree of specialization and then disappeared; but it is manifest they did do so. Still the fact of this early degree of excessive specialization tells to a certain, however small, extent against a progress through excessively minute steps, whether fortuitous or not; as also does the distinctness of forms formerly supposed to constitute connecting links. For, it must not be forgotten, that if species have manifested themselves generally by gradual and minute modifications, then the absence, not in one but in *all cases,* of such connecting links, is a phenomenon which remains to be accounted for.

It appears then that, apart from fortuitous changes, there are certain difficulties in the way of accepting extremely minute modifications of any kind, although these difficulties may not be insuperable. Something, at all events, is to be said in favour of the opinion that sudden and appreciable changes have from time to time occurred, however they may have been induced.   Marked *races* have undoubtedly so arisen (some striking instances having been here recorded), and it is at least conceivable that such may be the mode of *specific* manifestation generally, the possible conditions as to which will be considered in a later chapter.

CHAPTER V.

AS TO SPECIFIC STABILITY.

What is meant by the phrase "specific stability;" such stability to be
expected *a priori*, or else considerable changes at once.—Rapidly
increasing difficulty of intensifying race characters; alleged causes of
this phenomenon; probably an internal cause co-operates.—A certain
definiteness in variations.—Mr. Darwin admits the principle of specific
stability in certain cases of unequal variability.—The goose.—The
peacock —The guinea fowl.—Exceptional causes of variation under
domestication.—Alleged tendency to reversion.—Instances.—Sterility
of hybrids.—Prepotency of pollen of same species, but of different
race.—Mortality in young gallinaceous hybrids.—A bar to intermixture exists somewhere.—Guinea-pigs.—Summary and conclusion.

As was observed in the preceding chapters, arguments may yet
be advanced in favour of the opinion that species are stable
(at least in the intervals of their comparatively sudden successive
manifestations); that the organic world consists, according to
Mr. Galton's before-mentioned conception, of many facetted
spheroids, each of which can repose upon any one facet, but,
when too much disturbed, rolls over till it finds repose in stable
equilibrium upon another and distinct facet. Something, it is
here contended, may be urged, in favour of the existence of such
facets—of such intermitting conditions of stable equilibrium.

A view as to the stability of species, in the intervals of
change, has been well expressed in an able article, before quoted
from, as follows :[1]—" A given animal or plant appears to be con-

---

[1] *North British Review,* New Series, vol. vii., March 1867, p. 282.

tained, as it were, within a sphere of variation: one individual lies near one portion of the surface ; another individual, of the same species, near another part of the surface ; the average animal at the centre. Any individual may produce descendants varying in any direction, but is more likely to produce descendants varying towards the centre of the sphere, and the variations in that direction will be greater in amount than the variations towards the surface." This might be taken as the representation of the normal condition of species (*i.e.* during the periods of repose of the several facets of the spheroids), on that view which, as before said, may yet be defended.

Judging the organic world from the inorganic, we might expect, *a priori*, that each species of the former, like crystallized species, would have an approximate limit of form, and even of size, and at the same time that the organic, like the inorganic forms, would present modifications in correspondence with surrounding conditions ; but that these modifications would be, not minute and insignificant, but definite and appreciable, equivalent to the shifting of the spheroid on to another facet for support.

Mr. Murphy says,[1] "Crystalline formation is also dependent in a very remarkable way on the medium in which it takes place." "Beudant has found that common salt crystallizing from pure water forms cubes, but if the water contains a little boracic acid, the angles of the cubes are truncated. And the Rev. E. Craig has found that carbonate of copper, crystallizing from a solution containing sulphuric acid, forms hexagonal tubular prisms ; but if a little ammonia is added, the form changes to that of a long rectangular prism, with secondary planes in the angles. If a little more ammonia is added, several varieties of rhombic octahedra appear ; if a little nitric acid is added, the rectangular prism appears again. The changes take place not by the addition of new crystals, but by changing the growth of

1 "Habit and Intelligence," vol. i. p. 75.

the original ones." These, however, may be said to be the same species, after all ; but recent researches by Dr. H. Charlton-Bastian seem to show that modifications in the conditions may result in the evolution of forms so diverse as to constitute different organic species.

Mr. Murphy observes [1] that "it is scarcely possible to doubt that the various forms of fungi which are characteristic of particular situations are not really distinct species, but that the same germ will develop into different forms, according to the soil on which it falls ; " but it is possible to interpret the facts differently, and it may be that these are the manifestations of really different and distinct species, developed according to the different and distinct circumstances in which each is placed. Mr. Murphy quotes Dr. Carpenter [2] to the effect that " No *Puccinia* but the *Puccinia rosœ* is found upon rose bushes, and this is seen nowhere else ; *Omygena exigua* is said to be never seen but on the hoof of a dead horse ; and *Isaria felina* has only been observed upon the dung of cats, deposited in humid and obscure situations." He adds, " We can scarcely believe that the air is full of the germs of distinct species of fungi, of which one never vegetates until it falls on the hoof of a dead horse, and another till it falls on cat's dung in a damp and dark place." This is true, but it does not quite follow that they are necessarily the same species if, as Dr. Bastian seems to show, thoroughly different and distinct organic forms [3] can be evolved one from another by modifying the conditions. This observer has brought forward arguments and facts from which it would appear that such definite, sudden. and considerable transformations may take place in the lowest organisms. If such is really the case, we might expect, *a priori*, to find in the highest organisms a tendency (much more impeded

[1] "Habit and Intelligence," vol. i. p. 202.

[2] " Comparative Physiology," p. 214, note.

[3] See *Nature*, June and July 1870, Nos. 35, 36, and 37, pp. 170, 193, and 219.

and rare in its manifestations) to similarly appreciable and sudden changes, under certain stimuli; but a tendency to continued stability, under normal and ordinary conditions. The proposition that species have, under ordinary circumstances, a definite limit to their variability, is largely supported by facts brought forward by the zealous industry of Mr. Darwin himself. It is unquestionable that the degrees of variation which have been arrived at in domestic animals have been obtained more or less readily in a moderate amount of time, but that further development in certain desired directions is in some a matter of extreme difficulty, and in others appears to be all but, if not quite, an impossibility. It is also unquestionable that the degree of divergence which has been attained in one domestic species is no criterion of the amount of divergence which has been attained in another. It is contended on the other side that we have no evidence of any limits to variation other than those imposed by physical conditions, such, *e.g.*, as those which determine the greatest degree of speed possible to any animal (of a given size) moving over the earth's surface; also it is said that the differences in degree of change shown by different domestic animals depend in great measure upon the abundance or scarcity of individuals subjected to man's selection, together with the varying direction and amount of his attention in different cases; finally, it is said that the changes found in nature are within the limits to which the variation of domestic animals extends,—it being the case that when changes of a certain amount have occurred to a species under nature, it becomes *another species*, or sometimes *two or more other species* by divergent variations, each of these species being able again to vary and diverge in any useful direction.

But the fact of the rapidly increasing difficulty found in producing by ever such careful selection, any further extreme in some change already carried very far (such as the tail of the "fan-tailed pigeon" or the crop of the "pouter"), is certainly, so

far as it goes, on the side of the existence of definite limits to
vaiiability.   It is asserted in reply, that physiological conditions
of health and life may bar any such further development. Thus,
Mr. Wallace says[1] of these developments : "Variation s ems to
have reached its limits in these birds.   But so it has in nature.
The fantail has not only more tail-feathers than any of the three
hundred and forty existing species of pigeons, but more than
any of the eight thousand known species of birds.   There is, of
course, some limit to the number of feathers of which a tail
useful for flight can consist, and in the fantail we have probably
reached that limit.   Many birds have the œsophagus or the
skin of the neck more or less dilatable, but in no known bird is
it so dilatable as in the pouter pigeon.   Here again the possible limit, compatible with a healthy existence, has probably
been reached.   In like manner, the differences in the size and
form of the beak in the various breeds of the domestic pigeon,
is greater than that between the extreme forms of beak in the
various genera and sub-families of the whole pigeon tribe. From
these facts, and many others of the same nature, we may fairly
infer, that if rigid selection were applied to any organ, we cou'd
in a comparatively short time produce a much greater amount of
change than that which occurs between species and species in a
state of nature, since the differences which we do produce are
often comparable with those which exist between distinct genera
or distinct families."

    But in a domestic bird like the fantail where Natural Selec-
tion does not come into play, the tail-feathers could hardly be
limited by "utility for flight," yet two more tail-feathers could
certainly exist in a fancy breed if "utility for flight" were the
only obstacle.   It seems probable that the real barrier is an
internal one in the nature of the organism, and the existence of
such is just what is contended for in this chapter.   As to the

                    [1] "Natural Selection," p. 293.

differences between domestic races being greater than those between species or even genera, that is not enough for the argument. For upon the theory of "Natural Selection" all birds have a common origin, from which they diverged by infinitesimal changes, so that we ought to meet with sufficient changes to warrant the belief that a hornbill could be produced from a humming-bird, proportionate time being allowed.

But not only does it appear that there are barriers which oppose change in certain directions, but that there are positive tendencies to development along certain special lines. In a bird which has been kept and studied like the pigeon, it is difficult to believe that any remarkable spontaneous variations would pass unnoticed by breeders, or that they would fail to be attended to and developed by some one fancier or other. On the hypothesis of *indefinite* variability, it is then hard to say why pigeons with bills like toucans, or with certain feathers lengthened like those of trogans, or those of birds of paradise, have never been produced. This, however, is a question which may be settled by experiment. Let a pigeon be bred with a bill like a toucan's, and with the two middle tail-feathers lengthened like those of the king bird of paradise, or even let individuals be produced which exhibit any marked tendency of the kind, and indefinite variability shall be at once conceded.

As yet all the changes which have taken place in pigeons are of a few definite kinds only, such as may be well conceived t be compatible with a species possessed of a certain inherent capacity for considerable yet definite variation, a capacity for the ready production of certain degrees of abnormality, which then cannot be further increased.

Mr. Darwin himself has already acquiesced in the proposition here maintained, inasmuch as he distinctly affirms the existence of a marked internal barrier to change in certain cases. And if this is admitted in one case, the *principle* is conceded, and it immediately becomes probable that such internal barriers exist

in all, although enclosing a much larger field for variation in
some cases than in others.   Mr. Darwin abundantly demonstrates the variability of dogs, horses, fowls, and pigeons, but he
none the less shows clearly the *very small* extent to which the
goose, the peacock, and the guinea-fowl have varied.[1]   Mr.
Darwin attempts to explain this fact as regards the goose by the
animal being valued only for food and feathers, and from no
pleasure having been felt in it on other accounts.   He adds,
however, at the end the striking remark,[2] which concedes the
whole position, " but the goose seems to have a *singularly inflexible organization.*"   This is not the only place in which such
expressions are used.   He elsewhere makes use of phrases which
quite harmonize with the conception of a normal specific constancy, but varying greatly and suddenly at intervals.   Thus he
speaks[3] of a *whole organization seeming to have become plastic,
and tending to depart from the parental type.*   That different
organisms should have different degrees of variability, is only
what might have been expected *a priori* from the existence of
parallel differences in inorganic species, some of these having but
a single form, and others being polymorphic.

To return to the goose, however, it may be remarked that it
is at least as probable that its fixity of character is the cause of
the neglect, as the reverse.   It is by no means unfair to assume
that *had* the goose shown a tendency to vary similar in degree
to the tendency to variation of the fowl or pigeon, it would
have received attention at once on that account.

As to the peacock it is excused on the pleas (1), that the individuals maintained are so few in number, and (2) that its beauty
is so great it can hardly be improved.   But the individuals
maintained *have not been too few* for the independent origin of
the black-shouldered form, or for the supplanting of the

[1] "Animals and Plants under Domestication," vol. i. pp. 289—295.
[2] "Origin of Species," 5th edition, 1869, p. 45.
[3] Ibid. p. 13.

commoner one by it. As to any neglect in selection, it can hardly be imagined that with regard to this bird (kept as it is all but exclusively for its beauty), any spontaneous beautiful variation in colour or form would have been neglected. On the contrary, it would have been seized upon with avidity and preserved with anxious care. Yet apart from the black-shouldered and white varieties, no tendency to change has been known to show itself. As to its being too beautiful for improvement, that is a proposition which can hardly be maintained. Many consider the Javan bird as much handsomer than the common peacock, and it would be easy to suggest a score of improvements as regards either species.

The guinea-fowl is excused, as being " no general favourite, and scarcely more common than the peacock ;" but Mr. Darwin himself shows and admits that it is a noteworthy instance of constancy under very varied conditions.

These instances alone (and there are yet others) seem sufficient to establish the assertion, that degree of change is different in different domestic animals. It is, then, somewhat unwarrantable in any Darwinian to assume that *all* wild animals have a capacity for change similar to that existing in *some* of the domestic ones. It seems more reasonable to assert the opposite, namely, that if, as Mr. Darwin says, the capacity for change is different in different domestic animals, it must surely be limited in those which have it least, and *a fortiori* limited in wild animals.

Indeed, it cannot be reasonably maintained that wild species certainly vary as much as do domestic races ; it is possible that they may do so, but at least this has not been yet shown. Indeed, the much greater degree of variation amongst domestic animals than amongst wild ones is asserted over and over again by Mr. Darwin, and his assertions are supported by an overwhelming mass of facts and instances.

Of course, it may be asserted that a tendency to indefinite change exists in all cases, and that it is only the circum-

stances and conditions of life which modify the effects of this
tendency to change so as to produce such different results in
different cases. But assertion is not proof, and this assertion
has not been proved. Indeed, it may be equally asserted (and
the statement is more consonant with some of the facts given),
that domestication in certain animals induces and occasions a
capacity for change which is wanting in wild animals—the introduction of new causes occasioning new effects. For, though a
certain degree of variability (normally, in all probability, only
oscillation) exists in all organisms, yet domestic ones are exposed to new and different causes of variability, resulting in such
striking divergencies as have been observed. Not even in this
latter case, however, is it necessary to believe that the variability
is indefinite, but only that the small oscillations become in certain
instances intensified into large and conspicuous ones. Moreover,
it is possible that some of our domestic animals have been in
part chosen and domesticated through possessing variability in
an eminent degree.

That each species exhibits certain oscillations of structure is
admitted on all hands. Mr. Darwin asserts that this is the
exhibition of a tendency to vary which is absolutely indefinite.
If this indefinite variability *does* exist, of course no more need
be said. But we have seen that there are arguments *a priori*
and *a posteriori* against it, while the occurrence of variations in
certain domestic animals greater in degree than the differences
between many wild species, is no argument in favour of its
existence, until it can be shown that the causes of variability in
the one case are the same as in the other. An argument against
it, however, may be drawn from the fact, that certain animals,
though placed under the influence of those exceptional causes of
variation to which domestic animals are subject, have yet never
been known to vary, even in a degree equal to that in which
certain wild kinds have been ascertained to vary.

In addition to this immutability of character in some animals, it is undeniable, that domestic varieties have little stability, and much tendency to reversion, whatever be the true explanation of such phenomena.

In controverting the generally received opinion as to "reversion," Mr. Darwin has shown that it is not all breeds which in a few years revert to the original forms; but he has shown no more. Thus, the feral rabbits of Porto Santo, Jamaica, and the Falkland Islands, have not yet so reverted in those several localities.[1] Nevertheless, a Porto Santo rabbit brought to England reverted in a manner the most striking, recovering the proper colour of its fur "in rather less than four years."[2] Again, the white silk fowl, in our climate, "reverts to the ordinary colour of the common fowl in its skin and bones, due care having been taken to prevent any cross."[3] This reversion taking place in spite of careful selection, is very remarkable.

Numerous other instances of reversion are given by Mr. Darwin, both as regards plants and animals; amongst others, the singular fact of bud reversion.[4] The curiously recurring development of black sheep, in spite of the most careful breeding, may also be mentioned, though, perhaps, reversion has no part in the phenomenon.

These facts seem certainly to tell in favour of limited variability, while the cases of non-reversion do not contradict it, as it is not contended that all species have the same tendency to revert, but rather that their capacities in this respect, as well as for change, are different in different kinds, so that often reversion may only show itself at the end of very long periods indeed.

Yet some of the instances given as probable or possible causes of reversion by Mr. Darwin, can hardly be such. He cites, for example, the occasional presence of supernumerary digits in man.[5] For this notion, however, he is not responsible,

[1] "Animals and Plants under Domestication," vol. i. p. 115.
[2] Ibid. vol. i. p. 114.   [3] Ibid. vol. i. p. 243.
[4] Ibid. vol. ii. p. 361.   [5] Ibid. vol. ii. p. 16.

as he rests his remark on the authority of a passage published
by Professor Owen. Again, he refers[1] to " the greater frequency
of a monster proboscis in the pig than in any other animal."
But with the exception of the peculiar muzzle of the Saiga (or
European antelope), the only known proboscidian Ungulates are
the elephants and tapirs, and to neither of these has the pig
any close affinity. It is rather in the horse than in the pig that
we might look for the appearance of a reversionary proboscis,
as both the elephants and the tapirs have the toes of the hind
foot of an odd number. It is true that the elephants are
generally considered to form a group apart from both the odd
and the even-toed Ungulata. But of the two, their affinities
with the odd-toed division are more marked.[2]

Another argument in favour of the, at least intermitting, constancy of specific forms and of sudden modification, may be
drawn from the absence of minute transitional forms, but this
will be considered in the next chapter.

It remains now to notice in favour of specific stability,
that the objection drawn from physiological difference between
" species " and " races " still exists unrefuted.

Mr. Darwin freely admits difficulties regarding the sterility
of different species when crossed, and shows satisfactorily
that it could never have arisen from the action of " Natural
Selection." He remarks[3] also: " With some few exceptions, in
the case of plants, domesticated varieties, such as those of the
dog, fowl, pigeon, several fruit trees, and culinary vegetables,
which differ from each other in external characters more than
many species, are perfectly fertile when crossed, or even fertile

---

[1] " Animals and Plants under Domestication," vol. ii. p. 57.

[2] This has been shown by my late friend, Mr. H. N. Turner, jun., in an
excellent paper by him in the " Proceedings of the Zoological Society for
1849," p. 147. The untimely death, through a dissecting wound, of this
most promising young naturalist, was a very great loss to zoological science.

[3] " Animals and Plants under Domestication," vol. ii. p. 189.

in excess, whilst closely allied species are almost invariably in some degree sterile."

Again, after speaking of "the general law of good being derived from the intercrossing of distinct individuals of the same species," and the evidence that the pollen of a distinct *variety* or race is prepotent over a flower's own pollen, adds the very significant remark,[1] " When distinct *species* are crossed, the case is directly the reverse, for a plant's own pollen is almost always prepotent over foreign pollen."

Again he adds:[2] " I believe from observations communicated to me by Mr. Hewitt, who has had great experience in hybridizing pheasants and fowls, that the early death of the embryo is a very frequent cause of sterility in first crosses. Mr. Salter has recently given the results of an examination of about 500 eggs produced from various crosses between three species of Gallus and their hybrids. The majority of these eggs had been fertilized, and in the majority of the fertilized eggs the embryos either had been partially developed and had then aborted, or had become nearly mature, but the young chickens had been unable to break through the shell. Of the chickens which were born, more than four-fifths died within the first few days, or at latest weeks, ' without any obvious cause, apparently from mere inability to live,' so that from 500 eggs only twelve chickens were reared. The early death of hybrid embryos probably occurs in like manner with plants, at least it is known that hybrids raised from very distinct species are sometimes weak and dwarfed, and perish at an early age, of which fact Max Wichura has recently given some striking cases with hybrid willows."

Mr. Darwin objects to the notion that there is any special sterility imposed to check specific intermixture and change, saying,[3] " To grant to species the special power of producing

---

[1] "Origin of Species," 5th edition, 1869, p. 115.
[2] Ibid. p. 322.   [3] Ibid. p. 314.

hybrids, and then to stop their further propagation by different degrees of sterility, not strictly related to the facility of the first union between their parents, seems a strange arrangement."

But this only amounts to saying that the author himself would not have so acted had he been the Creator. A "strange arrangement" must be admitted anyhow, and all who acknowledge teleology at all, must admit that the strange arrangement was designed. Mr. Darwin says, as to the sterility of species, that the cause lies exclusively in their sexual constitution; but all that need be affirmed is that sterility is brought about somehow, and it is undeniable that "crossing" *is* checked. All that is contended for is that there *is* a bar to the intermixture of *species*, but not of *breeds;* and if the conditions of the generative products are that bar, it is enough for the argument, no special kind of barring action being contended for.

He, however, attempts to account for the modification of the sexual products of species as compared with those of varieties, by the exposure of the former to more uniform conditions during longer periods of time than those to which varieties are exposed, and that as wild animals, when captured, are often rendered sterile by captivity, so the influence of union with another species may produce a similar effect. It seems to the author an unwarrantable assumption that a cross with what, on the Darwinian theory, can only be a slightly diverging descendant of a common parent, should produce an effect equal to that of captivity, and consequent change of habit, as well as considerable modification of food.

No clear case has been given by Mr. Darwin in which mongrel animals, descended from the same undoubted species, have been persistently infertile *inter se;* nor any clear case in which hybrids between animals, generally admitted to be distinct species, have been continuously fertile *inter se.*

It is true that facts are brought forward tending to establish the probability of the doctrine of Pallas, that species may sometimes be rendered fertile by domestication. But even if this were true, it would be no approximation towards proving the converse, *i.e.* that races and varieties may become sterile when wild. And whatever may be the preference occasionally shown by certain breeds to mate with their own variety, no sterility is recorded as resulting from unions with other varieties. Indeed, Mr. Darwin remarks,[1] " With respect to sterility from the crossing of domestic races, I know of no well-ascertained case with animals. This fact (seeing the great difference in structure between some breeds of pigeons, fowls, pigs, dogs, &c.) is extraordinary when contrasted with the sterility of many closely-allied natural species when crossed."

It has been alleged that the domestic and wild guinea-pig do not breed together, but the specific identity of these forms is very problematical. Mr. A. D. Bartlett, superintendent of the Zoological Gardens, whose experience is so great, and observation so quick, believes them to be decidedly distinct species.

Thus, then, it seems that a certain normal specific stability in species, accompanied by occasional sudden and considerable modifications, might be expected *a priori* from what we know of crystalline inorganic forms and from what we may anticipate with regard to the lowest organic ones. This presumption is strengthened by the knowledge of the increasing difficulties which beset any attempt to indefinitely intensify any race characteristics. The obstacles to this indefinite intensification, as well as to certain lines of variation in certain cases, appear to be not only external, but to depend on internal causes or an internal cause. We have seen that Mr. Darwin himself implicitly admits the principle of specific stability in asserting the singular inflexibility of the organization of the goose. We have also seen that it is not fair to conclude that all wild races can vary as much as the most variable domestic ones. It has

"Animals and Plants under Domestication," vol. ii. p. 104.

also been shown that there are grounds for believing in a tendency to reversion generally, as it is distinctly present in certain instances. Also that specific stability is confirmed by the physiological obstacles which oppose themselves to any considerable or continued intermixture of species, while no such barriers oppose themselves to the blending of varieties. All these considerations taken together may fairly be considered as strengthening the belief that specific manifestations are relatively stable. At the same time the view advocated in this book does not depend upon, and is not identified with, any such stability. All that the Author contends for is that specific manifestation takes place along certain lines, and according to law, and not in an exceedingly minute, indefinite, and fortuitous manner. Finally, he cannot but feel justified, from all that has been brought forward, in reiterating the opening assertion of this chapter that something is still to be said for the view which maintains that species are stable, at least in the intervals of their comparatively rapid successive manifestations.

CHAPTER VI.

SPECIES AND TIME.

Two relations of species to time.—No evidence of past existence of minutely
intermediate forms when such might be expected *a priori*.—Bats,
Pterodactyles, Dinosauria, and Birds.—Ichthyosauria, Chelonia, and
Anoura.—Horse ancestry.—Labyrinthodonts and Trilobites.—Two sub-
divisions of the second relation of species to time.—Sir William Thomson's views.—Probable period required for ultimate specific evolution
from primitive ancestral forms.—Geometrical increase of time required
for rapidly multiplying increase of structural differences.—Proboscis
monkey. —Time required for deposition of strata necessary for Darwinian evolution.—High organization of Silurian forms of life.—
Absence of fossils in oldest rocks.—Summary and conclusion.

Two considerations present themselves with regard to the necessary relation of species to time if the theory of "Natural Selection" is valid and sufficient.

The first is with regard to the evidences of the past existence
of intermediate forms, their duration and succession.

The second is with regard to the total amount of time
required for the evolution of all organic forms from a few
original ones, and the bearing of other sciences on this question
of time.

As to the first consideration, evidence is as yet against the
modification of species by "Natural Selection" alone, because
not only are minutely transitional forms generally absent, but
they are absent in cases where we might certainly *a priori*
have expected them to be present.

Now it has been said:[1] "If Mr. Darwin's theory be true, the number of varieties differing one from another a very little must have been indefinitely great, so great indeed as probably far to exceed the number of individuals which have existed of any one variety. If this be true, it would be more probable that no two specimens preserved as fossils should be of one variety than that we should find a great many specimens collected from a very few varieties, provided, of course, the chances of preservation are equal for all individuals." "It is really strange that vast numbers of perfectly similar specimens should be found, the chances against their perpetuation as fossils are so great; but it is also very strange that the specimens should be so exactly alike as they are, if, in fact, they came and vanished by a gradual change."

Mr. Darwin attempts[2] to show cause why we should believe *a priori* that intermediate varieties would exist in lesser numbers than the more extreme forms; but though they would doubtless do so sometimes, it seems too much to assert that they would do so generally, still less universally. Now little less than universal and very marked inferiority in numbers would account for the absence of certain series of minutely intermediate fossil specimens. The mass of palæontological evidence is indeed overwhelmingly against minute and gradual modification. It is true that when once an animal has obtained powers of flight its means of diffusion are indefinitely increased, and we might expect to find many relics of an aërial form and few of its antecedent state—with nascent wings just commencing their suspensory power. Yet had such a slow mode of origin, as Darwinians contend for, operated exclusively in all cases, it is absolutely incredible that birds, bats, and pterodactyles should have left the remains they have, and yet not a

---

[1] *North British Review*, New Series, vol. vii., March 1867, p. 317.
[2] "Origin of Species," 5th edition, 1869, p. 212.

single relic be preserved in any one instance of any of these different forms of wing in their incipient and relatively imperfect functional condition !

WING-BONES OF PTERODACTYLE, BAT, AND BIRD.

Whenever the remains of bats have been found they have presented the exact type of existing forms, and there is as yet no indication of the conditions of an incipient elevation from the ground.

The pterodactyles, again, though a numerous group, are all true and perfect pterodactyles, though surely *some* of the many incipient forms, which on the Darwinian theory have existed, must have had a good chance of preservation.

As to birds, the only notable instance in which discoveries recently made appear to fill up an important hiatus, is the interpretation given by Professor Huxley [1] to the remains of Dinosaurian reptiles, and which were noticed in the third chapter of this work. The learned Professor has (as also has Professor Cope in America) shown that in very important and

[1] See also the *Popular Science Review* for July 1868.

significant points the skeletons of the Iguanodon and of its allies approach very closely to that existing in the ostrich, emeu, rhea, &c. He has given weighty reasons for thinking that the line of affinity between birds and reptiles passes to the birds last named from the Dinosauria rather than from the Pterodactyles, through Archeopteryx-like forms to the ordinary birds. Finally, he has thrown out the suggestion that the celebrated footsteps left by some extinct three-toed creatures on the very ancient sandstone of Connecticut were made, not, as hitherto supposed, by true birds, but by more or less ornithic reptiles. But even supposing all that is asserted or inferred on this subject to be fully proved, it would not approach to a demonstration of specific origin by *minute* modification. And though it harmonizes well with "Natural Selection," it is equally consistent with the rapid and sudden development of new specific forms of life. Indeed, Professor Huxley, with a laudable caution and moderation too little observed by some Teutonic Darwinians, guarded himself carefully from any imputation of asserting dogmatically the theory of "Natural Selection," while upholding fully the doctrine of evolution.

But, after all, it is by no means certain, though very probable, that the Connecticut footsteps were made by very ornithic reptiles, or extremely sauroid birds. And it must not be forgotten that a completely carinate[1] bird (the Archeopteryx) existed at a time, when, as yet, we have no evidence of some of the Dinosauria having come into being. Moreover, if the remarkable and minute similarity of the coracoid of a pterodactyle to that of a bird be merely the result of function and no sign of genetic affinity, it is not inconceivable that pelvic and leg resemblances of Dinosauria to birds may be functional likewise,

---

[1] A bird with a keeled breast-bone, such as almost all existing birds possess.

though such an explanation is, of course, by no means necessary
to support the view maintained in this book.

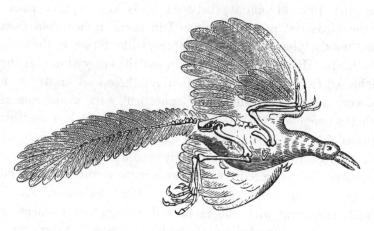

THE ARCHEOPTERYX (OF THE OOLITE STRATA).

But the number of forms represented by many individuals,
yet by *no transitional ones*, is so great that only two or three
can be selected as examples. Thus those remarkable fossil
reptiles, the Ichthyosauria and Plesiosauria, extended, through the

SKELETON OF AN ICHTHYOSAURUS.

secondary period, probably over the greater part of the globe.
Yet no single transitional form has yet been met with in spite
of the multitudinous individuals preserved. Again, with their
modern representatives the Cetacea, one or two aberrant forms

alone have been found, but no series of transitional ones indi-
cating minutely the line of descent.   This group, the whales,
is a very marked one, and it is curious, on Darwinian principles,

SKELETON OF A PLESIOSAURUS.

that so few instances tending to indicate its mode of origin
should have presented themselves.   Here, as in the bats, we
might surely expect that some relics of unquestionably incipient
stages of its development would have been left.

The singular order Chelonia, including the tortoises, turtles,
and terrapins (or fresh-water tortoises), is another instance of
an extreme form without any, as yet known, transitional stages.
Another group may be finally mentioned, viz. the frogs and
toads, anourous Batrachians, of which we have at present no
relic of any kind linking them on to the Eft group on the one
hand, or to reptiles on the other.

The only instance in which an approach towards a series of
nearly related forms has been obtained is the existing horse, its
predecessor Hipparion and other extinct forms.   But even here
there is no proof whatever of modification by minute and
infinitesimal steps ; *a fortiori* no approach to a proof of
modification by " Natural Selection," acting upon indefinite
fortuitous variations.   On the contrary, the series is an admirable example of successive modification in one special
direction along one beneficial line, and the teleologist must

here be allowed to consider that one motive of this modification (among probably an indefinite number of motives inconceivable to us) was the relationship in which the horse was to stand to the human inhabitants of this planet. These extinct forms, as Professor Owen remarks,[1] "differ from each other in a greater degree than do the horse, zebra, and ass," which are not only good *zoological* species as to form, but are species *physiologically*, *i.e.* they cannot produce a race of hybrids fertile *inter se.*

As to the mere action of surrounding conditions, the same Professor remarks:[2] "Any modification affecting the density of the soil might so far relate to the changes of limb-structure, as that a foot with a pair of small hoofs dangling by the sides of the large one, like those behind the cloven hoof of the ox, would cause the foot of Hipparion, *e.g.*, and *a fortiori* the broader based three-hoofed foot of the Palæothere, to sink less deeply into swampy soil, and be more easily withdrawn than the more concentratively simplified and specialized foot of the horse. Rhinoceroses and zebras, however, tread together the arid plains of Africa in the present day ; and the horse has multiplied in that half of America where two or more kinds of tapir still exist. That the continents of the Eocene or Miocene periods were less diversified in respect of swamp and sward, pampas or desert, than those of the Pliocene period, has no support from observation or analogy."

Not only, however, do we fail to find any traces of the incipient stages of numerous very peculiar groups of animals, but it is undeniable that there are instances which appeared at first to indicate a *gradual transition*, yet which instances have been shown by further investigation and discovery not to indicate truly anything of the kind. Thus at one time the remains of Labyrinthodonts, which up till then had been discovered, seemed to justify the opinion that as time went on, forms had

---

[1] "Anatomy of Vertebrates," vol. iii. p. 792.   [2] Ibid. p. 793.

successively appeared with more and more complete segmentation and ossification of the backbone, which in the earliest forms was (as it is in the lowest fishes now) a soft continuous rod or

TRILOBITE.

notochord. Now, however, it is considered probable that the soft back-boned Labyrinthodont, Archegosaurus, was an immature or larval form,[1] while Labyrinthodonts with completely developed vertebræ have been found to exist amongst the very earliest forms yet discovered. The same may be said regarding the eyes of the trilobites, some of the oldest forms having been found as well furnished in that respect as the very last of the group which has left its remains accessible to observation.

Such instances, however, as well as the way in which marked and special forms (as the Pterodactyles, &c., before referred to) appear at once in and similarly disappear from the geological record, are of course explicable on the Darwinian theory, provided a sufficiently enormous amount of past time be allowed. The alleged extreme, and probably great, imperfection of that record may indeed be pleaded in excuse. But it *is* an excuse.[2]

[1] As a tadpole is the *larval form* of a frog.

[2] As Professor Huxley, with his characteristic candour, fully admitted in his lecture on the Dinosauria before referred to.

Nor is it possible to deny the *a priori* probability of the preservation of at least a few *minutely transitional* forms in some instances if *every* species without exception has arisen exclusively by such minute and gradual transitions.

It remains, then, to turn to the other considerations with regard to the relation of species to time : namely (1) as to the total amount of time allowable by other sciences for organic evolution ; and (2) the proportion existing, on Darwinian principles, between the time anterior to the earlier fossils, and the time since ; as evidenced by the proportion between the amount of evolutionary change during the latter epoch and that which must have occurred anteriorly.

Sir William Thomson has lately[1] advanced arguments from three distinct lines of inquiry, and agreeing in one approximate result. The three lines of inquiry were—1. The action of the tides upon the earth's rotation. 2. The probable length of time during which the sun has illuminated this planet ; and 3. The temperature of the interior of the earth. The result arrived at by these investigations is a conclusion that the existing state of things on the earth, life on the earth, all geological history showing continuity of life, must be limited within some such period of past time as one hundred million years. The first question which suggests itself, supposing Sir W. Thomson's views to be correct, is, Is this period anything like enough for the evolution of all organic forms by "Natural Selection" ? The second is, Is this period anything like enough for the deposition of the strata which must have been deposited if all organic forms have been evolved by *minute* steps, according to the Darwinian theory ?

In the first place, as to Sir William Thomson's views, the Author of this book cannot presume to advance any opinion ; but the fact that they have not been refuted, pleads strongly in

---

[1] "Transactions of the Geological Society of Glasgow," vol. iii.

their favour when we consider how much they tell against the theory of Mr. Darwin. The last-named author only remarks that "many of the elements in the calculation are more or less doubtful,"[1] and Professor Huxley[2] does not attempt to *refute* Sir W. Thomson's arguments, but only to show cause for suspense of judgment, inasmuch as the facts *may be* capable of other explanations.

Mr. Wallace, on the other hand,[3] seems more disposed to accept them, and, after considering Sir William's objections and those of Mr. Croll, puts the probable date of the beginning of the Cambrian deposits[4] at only twenty-four million years ago. On the other hand, he seems to consider that specific change has been more rapid than generally supposed, and exceptionally stable during the last score or so of thousand years.

Now, first, with regard to the time required for the evolution of all organic forms by merely accidental, minute, and fortuitous variations, the useful ones of which have been preserved :

Mr. Murphy[5] is distinctly of opinion that there has not been time enough. He says, "I am inclined to think that geological time is too short for the evolution of the higher forms of life out of the lower by that accumulation of imperceptibly slow variations, to which alone Darwin ascribes the whole process."

"Darwin justly mentions the greyhound as being equal to any natural species in the perfect co-ordination of its parts, ' all adapted for extreme fleetness and for running down weak prey.' "

"Yet it is an artificial species (and not *physiologically* a species *at all*), formed by long-continued selection under domestication ; and there is no reason to suppose that any of the variations which have been selected to form it have been other than gradual and

---

[1] "Origin of Species," 5th edition, p. 354.
[2] See his address to the Geological Society, on February 19, 1869.
[3] See *Nature*, vol. i. p. 399, February 17, 1870.
[4] Ibid. vol. i. p. 454.
[5] " Habit and Intelligence," vol. i. p. 344.

almost imperceptible. Suppose that it has taken five hundred
years to form the greyhound out of his wolf-like ancestor. This
is a mere guess, but it gives the order of the magnitude." Now, if
so, "how long would it take to obtain an elephant from a pro-
tozoon, or even from a tadpole-like fish? Ought it not to take
much more than a million times as long?"[1]

Mr. Darwin[2] would compare with the natural origin of a
species "unconscious selection, that is, the preservation of the
most useful or beautiful animals, with no intention of modifying
the breed." He adds : " But by this process of unconscious
selection, various breeds have been sensibly changed in the
course of two or three centuries."

"Sensibly changed!" but not formed into "new species." Mr.
Darwin, of course, could not mean that species *generally* change
so rapidly, which would be strangely at variance with the abundant evidence we have of the stability of animal forms as represented on Egyptian monuments and as shown by recent deposits.
Indeed, he goes on to say,—" Species, however, probably change
much more slowly, and within the same country only a few
change at the same time. This slowness follows from all the
inhabitants of the same country being already so well adapted
to each other, that places in the polity of nature do not occur
until after long intervals, when changes of some kind in the
physical conditions, or through immigration, have occurred, and
individual differences and variations of the right nature, by
which some of the inhabitants might be better fitted to their
new places under altered circumstances, might not at once
occur." This is true, and not only will these changes occur at
distant intervals, but it must be borne in mind that in tracing
back an animal to a remote ancestry, we pass through modifications of such rapidly increasing number and importance that a
geometrical progression can alone indicate the increase of periods

[1] " Habit and Intelligence," vol. i. p. 345.
[2] "Origin of Species," 5th edition, p. 353.

which such profound alterations would require for their evolution through "Natural Selection" only.

Thus let us take for an example the proboscis monkey of Borneo (*Semnopithecus nasalis*). According to Mr. Darwin's own opinion, this form might have been " sensibly changed" in the course of two or three centuries. According to this, to evolve it as a true and perfect species one thousand years would be a very moderate period. Let ten thousand years be taken to represent approximately the period of substantially constant conditions during which no considerable change would be brought about. Now, if one thousand years may represent the period required for the evolution of the species *S. nasalis*, and of the other species of the genus Semnopithecus ; ten times that period should, I think, be allowed for the differentiation of that genus, the African Cercopithecus and the other genera of the family Simiidæ—the differences between the genera being certainly more than tenfold greater than those between the species of the same genus. Again we may perhaps interpose a period of ten thousand years' comparative repose.

For the differentiation of the families Simiidæ and Cebidæ— so very much more distinct and different than any two genera of either family—a period ten times greater should, I believe, be allowed than that required for the evolution of the subordinate groups. A similarly increasing ratio should be granted for the successive developments of the difference between the Lemuroid and the higher forms of primates ; for those between the original primate and other root-forms of placental mammals ; for those between primary placental and implacental mammals, and perhaps also for the divergence of the most ancient stock of these and of the monotremes, for in all these cases modifications of structure appear to increase in complexity in at least that ratio. Finally, a vast period must be granted for the development of the lowest mammalian type from the primitive stock of the whole vertebrate sub-kingdom. Supposing this primitive stock to have arisen directly from a very lowly organized animal indeed (such as a nematoid worm, or an ascidian, or a jelly-fish), yet it is not easy to believe that less than two thousand million years would be required for the totality of animal development by no other means than minute, fortuitous, occasional, and intermitting variations in all conceivable directions. If this be even an approximation to the truth, then there seem to be strong reasons for believing that geological time is not sufficient for such a process.

The second question is, whether there has been time enough for the deposition of the strata which must have been deposited, if all organic forms have been evolved according to the Darwinian theory?

Now this may at first seem a question for geologists only, but, in fact, in this matter geology must in some respects rather take its time from zoology than the reverse; for if Mr. Darwin's theory be true, past time down to the deposition of the Upper Silurian strata can have been but a very small fraction of that during which strata have been deposited. For when those Upper Silurian strata were formed, organic evolution had already run a great part of its course, perhaps the longest, slowest, and most difficult part of that course.

At that ancient epoch not only were the vertebrate, molluscous, and arthropod types distinctly and clearly differentiated, but highly developed forms had been produced in each of these sub-kingdoms. Thus in the Vertebrata there were fishes not belonging to the lowest but to the very highest groups which are known to have ever been developed, namely, the Elasmobranchs (the highly organized sharks and rays) and the Ganoids, a group now poorly represented, but for which the sturgeon may stand as a type, and which in many important respects more nearly resemble higher Vertebrata than do the ordinary or osseous fishes. Fishes in which the ventral fins are placed in front of the pectoral ones (*i.e.* jugular fishes) have been generally considered to be comparatively modern forms. But Professor

Huxley has kindly informed me that he has discovered a jugular fish in the Permian deposits.

Amongst the molluscous animals we have members of the very highest known class, namely, the Cephalopods, or cuttle-fish class; and amongst articulated animals we find Trilobites

A   B

CUTTLE-FISH.
A. Ventral aspect.   B. Dorsal aspect.

and Eurypterida, which do not belong to any incipient worm-like group, but are distinctly differentiated Crustacea of no low form.

We have in all these animal types nervous systems differ-entiated on distinctly different patterns, fully formed organs of circulation, digestion, excretion, and generation, complexly con-structed eyes and other sense organs; in fact, all the most elaborate and complete animal structures built up, and not only once, for in the fishes and mollusca we have (as described in the third chapter of this work) the coincidence of the independently developed organs of sense attaining a nearly similar complexity in two quite distinct forms. If, then, so small an advance has been made in fishes, molluscs, and arthropods since the Upper Silurian deposits, it will probably be within the mark to consider that the period before those deposits (during which all these organs would, on the Darwinian theory, have slowly built up their different perfections and complexities) occupied time at least a hundredfold greater.

Now it will be a moderate computation to allow 25,000,000 years for the deposition of the strata down to and including the Upper Silurian. If, then, the evolutionary work done during this deposition, only represents a hundredth part of the sum total, we shall require 2,500,000,000 (two thousand five hundred million) years for the complete development of the whole animal kingdom to its present state. Even one quarter of this, however, would far exceed the time which physics and astronomy seem able to allow for the completion of the process.

Finally, a difficulty exists as to the reason of the absence of rich fossiliferous deposits in the oldest strata—if life was then as abundant and varied as, on the Darwinian theory, it must have been. Mr. Darwin himself admits[1] "the case at present must remain inexplicable; and may be truly urged as a valid argument against the views" entertained in his book.

Thus, then, we find a wonderful (and on Darwinian principles an all but inexplicable) absence of minutely transitional forms. All the most marked groups, bats, pterodactyles, chelonians, ichthyosauria, anoura, &c., appear at once upon the scene. Even the horse, the animal whose pedigree has been probably best preserved, affords no conclusive evidence of specific origin by infinitesimal, fortuitous variations; while some forms, as the labyrinthodonts and trilobites, which seemed to exhibit gradual change, are shown by further investigation to do nothing of the sort. As regards the time required for evolution (whether estimated by

---

[1] "Origin of Species," 5th edition, p. 381.

the probably minimum period required for organic change or for the deposition· of strata which accompanied that change), reasons have been suggested why it is likely that the past history of the earth does not supply us with enough. First, because of the prodigious increase in the importance and number of differences and modifications which we meet with as we traverse successively greater and more primary zoölogical groups; and, secondly, because of the vast series of strata necessarily deposited if the period since the Lower Silurian marks but a small fraction of the period of organic evolution. Finally, the absence or rarity of fossils in the oldest rocks is a point at present inexplicable, and not to be forgotten or neglected.

Now all these difficulties are avoided if we admit that new forms of animal life of all degrees of complexity appear from time to time with comparative suddenness, being evolved according to laws in part depending on surrounding conditions, in part internal—similar to the way in which crystals (and, perhaps from recent researches, the lowest forms of life) build themselves up according to the internal laws of their component substance, and in harmony and correspondence with all environing influences and conditions.

CHAPTER VII.

SPECIES AND SPACE.

The geographical distribution of animals presents difficulties.—These not insurmountable in themselves; harmonize with other difficulties.—Fresh-water fishes.—Forms common to Africa and India; to Africa and South America; to China and Australia; to North America and China; to New Zealand and South America; to South America and Tasmania; to South America and Australia.—Pleurodont lizards.—Insectivorous mammals.—Similarity of European and South American frogs—Analogy between European salmon and fishes of New Zealand, &c. An ancient Antarctic continent probable.—Other modes of accounting for facts of distribution.—Independent origin of closely similar forms.—Conclusion.

THE study of the distribution of animals over the earth's surface presents us with many facts having certain not unimportant bearings on the question of specific origin. Amongst these are instances which, at least at first sight, appear to conflict with the Darwinian theory of "Natural Selection." It is not, however, here contended that such facts do by any means constitute by themselves obstacles which cannot be got over. Indeed it would be difficult to imagine any obstacles of the kind which could not be surmounted by an indefinite number of terrestrial modifications of surface — submergences and emergences—junctions and separations of continents in all directions and combinations of any desired degree of frequency. All this being supplemented by the intercalation of armies of enemies, multitudes of ancestors of all kinds, and myriads of connecting forms, whose *raison d'être* may be simply their utility or necessity for the support of the theory of " Natural Selection."

Nevertheless, when brought in merely to supplement and accentuate considerations and arguments derived from other sources, in that case difficulties connected with the geographical distribution of animals are not without significance, and are worthy of mention even though, by themselves, they constitute but feeble and more or less easily explicable puzzles which could not alone suffice either to sustain or to defeat any theory of specific origination.

Many facts as to the present distribution of animal life over the world are very readily explicable by the hypothesis of slight elevations and depressions of larger and smaller parts of its sur-·face, but there are others the existence of which it is much more difficult so to explain.

The distribution either of animals possessing the power of flight, or of inhabitants of the ocean, is, of course, easily to be accounted for ; the difficulty, if there is really any, must mainly be with strictly terrestrial animals of moderate or small powers of locomotion and with inhabitants of fresh water. Mr. Darwin himself observes,[1] " In regard to fish, I believe that the same species never occur in the fresh waters of distant continents." Now, the Author is enabled, by the labours and through the kindness of Dr. Günther, to show that this belief cannot be maintained ; he having been so obliging as to call attention to the following facts with regard to fish-distribution. These facts show that though only one species which is absolutely and exclusively an inhabitant of fresh water is as yet known to be found in distant continents, yet that in several other instances the same species *is* found in the fresh water of distant continents, and that very often the same *genus* is so distributed.

The genus *Mastacembelus* belongs to a family of fresh-water

---

[1] " Origin of Species," 5th edition, 1869, p. 463.

Indian fishes. Eight species of this genus are described by Dr. Günther in his catalogue.[1] These forms extend from Java and Borneo on the one hand, to Aleppo on the other. Nevertheless, a new species (*M. cryptacanthus*) has been described by the same author,[2] which is an inhabitant of the Camaroon country of *Western* Africa. He observes, "The occurrence of Indian forms on the West Coast of Africa, such as *Periophthalmus, Psettus, Mastacembelus*, is of the highest interest, and an almost new fact in our knowledge of the geographical distribution of fishes."

Ophiocephalus, again, is a truly Indian genus, there being no less than twenty-five species,[3] all from the fresh waters of the East Indies. Yet Dr. Günther informs me that there is a species in the Upper Nile and in West Africa.

The acanthopterygian family (*Labyrinthici*) contains nine fresh-water genera, and these are distributed between the East Indies and South and Central Africa.

The Carp fishes (Cyprinoids) are found in India, Africa, and Madagascar, but there are none in South America.

Thus existing fresh-water fishes point to an immediate connexion between Africa and India, harmonizing with what we learn from Miocene mammalian remains.

On the other hand, the Characinidæ (a family of the physostomous fishes) are found in Africa and South America, and not in India, and even its component groups are so distributed, —namely, the *Tetragonopterina* [4] and the *Hydrocyonina*.[5]

Again, we have similar phenomena in that almost exclusively fresh-water group the Siluroids.

Thus the genera *Clarias* [6] and *Heterobranchus* [7] are found both

---

[1] See his Catalogue of Acanthopterygian Fishes in the British Museum, vol. iii. p. 540.

[2] Proc. Zool. Soc. 1837, p. 102, and Ann. Mag. of Nat. Hist. vol. xx. p. 110.

[2] See Catalogue, vol. iii. p. 469.

[3] Ibid. vol. v. p. 311.

Ibid. p. 13.

[5] Ibid. p. 345.

[7] Ibid. p. 21.

in Africa and the East Indies. *Plotosus* is found in Africa, India, and Australia, and the species *P. anguillaris* [1] has been brought from both China and Moreton Bay. Here, therefore, we have the same species in two distinct geographical regions. It is however a coast fish, which, though entering rivers, yet lives in the sea.

Eutropius [2] is an African genus, but *E. obtusirostris* comes from India. On the other hand, *Amiurus* is a North American form; but one species, *A. cantonensis*,[3] comes from China.

The genus *Galaxias* [4] has at least one species common to New Zealand and South America, and one common to South America and Tasmania. In this genus we thus have an absolutely and completely fresh-water form *of the very same species* distributed between different and distinct geographical regions.

Of the lower fishes, a lamprey, *Mordacia mordax*,[5] is common to South Australia and Chile; while another form of the same family, namely, *Geotria chilensis*,[6] is found not only in South America and Australia, but in New Zealand also. These fishes, however, probably pass part of their lives in the sea.

We thus certainly have several species which *are* common to the fresh waters of distant continents, although it cannot be certainly affirmed that they are exclusively and entirely fresh-water fishes throughout all their lives except in the case of *Galaxias.*

Existing forms point to a close union between South America and Africa on the one hand, and between South America, Australia, Tasmania, and New Zealand on the other; but these unions were not synchronous any more than the unions indicated between India and Australia, China and Australia, China and North America, and India and Africa.

Pleurodont lizards are such as have the teeth attached by

---

[1] See Catalogue, vol. v. p. 24.   [2] Ibid. p. 52.   [3] Ibid. p. 100.
[4] Ibid. vol. vi. 208.   [5] Ibid. vol. viii. p. 507.   [6] Ibid. p. 509.

their sides to the inner surface of the jaw, in contradistinction to acrodont lizards, which have the bases of their teeth anchylosed to the summit of the margin of the jaw.   Now pleurodont

INNER SIDE OF LOWER JAW OF PLEURODONT LIZARD.
(Showing the teeth attached to the inner surface of its side.)

iguanian lizards abound in the South American region; but nowhere else, and are not as yet known to inhabit any part of the present continent of Africa.   Yet pleurodont lizards, strange to say, are found in Madagascar.   This is the more remarkable, inasmuch as we have no evidence yet of the existence in Madagascar of fresh-water fishes common to Africa and South America.

Again, that remarkable island Madagascar is the home of very singular and special insectivorous beasts of the genera Centetes, Ericulus, and Echinops; while the only other member of the group to which they belong is Solenodon, which is a resident in the West Indian Islands, Cuba and Hayti.   The connexion, however, between the West Indies and Madagascar must surely have been at a time when the great lemurine group was absent; for it is difficult to understand the spread of such a form as Solenodon, and at the same time the non-extension of the active lemurs, or their utter extirpation, in such a congenial locality as the West Indian Archipelago.

The close connexion of South America and Australia is demonstrated (on the Darwinian theory), not only from the marsupial fauna of both, but also from the frogs and toads which respectively inhabit those regions.   A truly remarkable similarity and parallelism exist, however, between certain of the same animals inhabiting South Western America and Europe. Thus Dr. Günther has described[1] a frog from Chile by the name of cacotus, which singularly resembles the European bombinator.

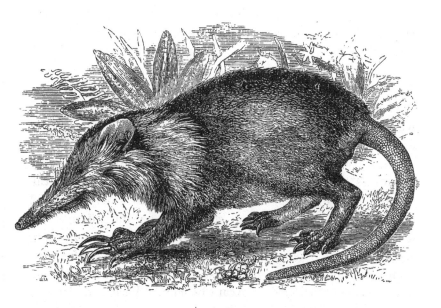

SOLENODON.

Again of the salmons, two genera from South America, New Zealand, and Australia, are analogous to European salmons.

In addition to this may be mentioned a quotation from Professor Dana, given by Mr. Darwin,[2] to the effect that "it is certainly a wonderful fact that New Zealand should have a closer resemblance in its crustacea to Great Britain, its antipode, than to any other part of the world:" and Mr. Darwin adds "Sir J. Richardson also speaks of the reappearance on the shores of New Zealand, Tasmania, &c. of northern forms of fish. Dr. Hooker informs me that twenty-five species of algæ are

[1] Proc. Zool. Soc. 1868, p. 482
[2] "Origin of Species," 5th edition, 1869, p. 454.

common to New Zealand and to Europe, but have not been found in the intermediate tropical seas."

Many more examples of the kind could easily be brought, but these must suffice. As to the last-mentioned cases Mr. Darwin explains them by the influence of the glacial epoch, which he would extend actually across the equator, and thus account, amongst other things, for the appearance in Chile of frogs having close genetic relations with European forms. But it is difficult to understand the persistence and preservation of such exceptional forms with the extirpation of all the others which probably accompanied them, if so great a migration of northern kinds had been occasioned by the glacial epoch.

Mr. Darwin candidly says,[1] "I am far from supposing that all difficulties in regard to the distribution and affinities of the identical and allied species, which now live so widely separated in the north and south, and sometimes on the intermediate mountain-ranges, are removed." ... "We cannot say why certain species and not others have migrated; why certain species have been modified and have given rise to new forms, whilst others have remained unaltered." Again he adds, "Various difficulties also remain to be solved; for instance, the occurrence, as shown by Dr. Hooker, of the same plants at points so enormously remote as Kerguelen Land, New Zealand, and Fuegia; but icebergs, as suggested by Lyell, may have been concerned in their dispersal. The existence, at these and other distant points of the southern hemisphere, of species which, though distinct, belong to genera exclusively confined to the south, is a more remarkable case. Some of these species are so distinct that we cannot suppose that there has been time since the commencement of the last glacial period for their migration and subsequent modification to the necessary degree." Mr. Darwin goes on to account for these facts by the probable existence of a rich antarctic

---

[1] "Origin of Species," 5th edition, p. 459.

flora in a warm period anterior to the last glacial epoch. There are indeed many reasons for thinking that a southern continent, rich in living forms, once existed. One such reason is the way in which struthious birds are, or have been, distributed around the antarctic region : as the ostrich in Africa, the rhea in South America, the emeu in Australia, the apteryx, dinornis, &c. in New Zealand, the epiornis in Madagascar. Still the existence of such a land would not alone explain-the various geographical cross relations which have been given above. It would not, for example, account for the resemblance between the crustacea or fishes of New Zealand and of England. It would, however, go far to explain the identity (specific or generic) between fresh water and other forms now simultaneously existing in Australia and South America, or in either or both of these, and New Zealand.

Again, mutations of elevation small and gradual (but frequent and intermitting), through enormous periods of time—waves, as it were, of land rolling many times in many directions—might be made to explain many difficulties as to geographical distribution, and any cases that remained would probably be capable of explanation, as being isolated but allied animal forms, now separated indeed, but being merely remnants of extensive groups which, at an earlier period, were spread over the surface of the earth. Thus none of the facts here given are any serious difficulty to the doctrine of "evolution," but it is contended in this book that if other considerations render it improbable that the manifestation of the successive forms of life has been brought about by minute, indefinite, and fortuitous variations, then these facts as to geographical distribution intensify that improbability, and are so far worthy of attention.

All geographical difficulties of the kind would be evaded if we could concede the probability of the independent origin, in different localities, of the same organic forms in animals high in the scale of nature. Similar causes must produce similar results, and new reasons have been lately adduced for believing, as regards the *lowest organisms,* that the same forms can arise and manifest themselves independently. The difficulty as to higher animals is, however, much greater, as (on the theory of evolution) one acting force must always be the ancestral history in each case, and this force must always tend to go on acting in the same groove and direction in the future as it has in the past. So that it is difficult to conceive that individuals, the ancestral history of which is very different, can be acted upon by all influences, external and internal, in such diverse ways and proportions that the results (unequals being added to unequals) shall be equal and similar. Still, though highly improbable, this cannot be said to be impossible; and if there *is* an innate law of any kind helping to determine specific evolution, this may more or less, or entirely, neutralize or even reverse the effect of ancestral habit. Thus, it is quite conceivable that a pleurodont lizard might have arisen in Madagascar in perfect independence of the similarly-formed American lacertilia : just as certain teeth of carnivorous and insectivorous marsupial animals have been seen most closely to resemble those of carnivorous and insectivorous placental beasts ; just as, again, the paddles of the Cetacea resemble, in the fact of a multiplication in the number of the phalanges, the many-jointed feet of extinct marine reptiles, and as the beak of the cuttle-fish or of the tadpole resembles that of birds. We have already seen (in Chapter III.) that it is impossible, upon any hypothesis, to escape admitting the independent origins of closely similar forms. It may be that they are both more frequent and more important than is generally thought.

That closely similar structures may arise without a genetic relationship has been lately well urged by Mr. Ray Lankester.[1]

---

[1] See Ann. and Mag. of Nat. Hist., July 1870, p. 37.

He has brought this notion forward even as regards the bones of the skull in osseous fishes and in mammals. He has done so on the ground that the probable common ancestor of mammals and of osseous fishes was a vertebrate animal of so low a type that it could not be supposed to have possessed a skull differentiated into distinct bony elements—even if it was bony at all. If this was so, then the cranial bones must have had an independent origin in each class, and in this case we have the most strikingly harmonious and parallel results from independent actions. For the bones of the skull in an osseous fish are so closely conformed to those of a mammal, that " both types of skull exhibit many bones in common," though " in each type some of these bones acquire special arrangements and very different magnitudes."[1] And no investigator of homologies doubts that a considerable number of the bones which form the skull of any osseous fish are distinctly homologous with the cranial bones of man. The occipital, the parietal, and frontal, the bones which surround the internal ear, the vomer, the premaxilla, and the quadrate bones, may be given as examples. Now, if such close relations of homology can be brought about independently of any but the most remote genetic affinity, it would be rash to affirm dogmatically that there is any impossibility in the independent origin of such forms as centetes and solenodon, or of genetically distinct batrachians, as similar to each other as are some of the frogs of South America and of Europe. At the same time such phenomena must at present be considered as very improbable, from the action of ancestral habit, as before stated.

We have seen, then, that the geographical distribution of animals presents difficulties, though not insuperable ones, for the Darwinian hypothesis. If, however, other reasons against it appear of any weight—if, especially, there is reason to believe that geological

[1] Professor Huxley's Lectures on the Elements of Comp. Anat. p. 184.

time has not been sufficient for it, then it will be well to bear in mind the facts here enumerated. These facts, however, are not opposed to the doctrine of evolution ; and if it could be established that closely similar forms had really arisen in complete independence one of the other, they would rather tend to strengthen and to support that theory.

CHAPTER VIII.

HOMOLOGIES.

Animals made-up of parts mutually related in various ways.—What homology is.—Its various kinds.—Serial homology.—Lateral homology.—
Vertical homology.--Mr. Herbert Spencer's explanations.—An internal
power necessary, as shown by facts of comparative anatomy.—Of teratology.—M. St. Hilaire.—Professor Burt Wilder.—Foot-wings.—Facts
of pathology.—Mr. James Paget.—Dr. William Budd.—The existence
of such an internal power of individual development diminishes the
improbability of an analogous law of specific origination.

THAT concrete whole which is spoken of as "an individual"
(such, *e.g.*, as a bird or a lobster) is formed of a more or less
complex aggregation of parts which are actually (from whatever
cause or causes) grouped together in a harmonious interdepen-
dency, and which have a multitude of complex relations amongst
themselves.

The mind detects a certain number of these relations as it
contemplates the various component parts of an individual in one
or other direction—as it follows up different lines of thought.
These perceived relations, though subjective, *as relations*, have
nevertheless an objective foundation as real parts, or conditions
of parts, of real wholes; they are, therefore, true relations, such,
e.g., as those between the right and left hand, between the hand
and the foot, &c.

The component parts of each concrete whole have also a relation of resemblance to the parts of other concrete wholes, whether

of the same or of different kinds, as the resemblance between the hands of two men, or that between the hand of a man and the fore-paw of a cat.

Now, it is here contended that the relationships borne one to another by various component parts, imply the existence of some innate, internal condition, conveniently spoken of as a power or tendency, which is quite as mysterious as is any innate condition, power, or tendency, resulting in the orderly evolution of successive specific manifestations. These relationships, as also this developmental power, will doubtless, in a certain sense, be somewhat further explained as science advances. But the result will be merely a shifting of the inexplicability a point backwards, by the intercalation of another step between the action of the internal condition or power and its external result. In the meantime, even if by " Natural Selection" we could eliminate the puzzles of the "origin of species," yet other phenomena, not less remarkable (namely, those noticed in this chapter), would still remain unexplained and as yet inexplicable. It is not improbable that, could we arrive at the causes conditioning all the complex inter-relations between the several parts of one animal, we should at the same time obtain the key to unlock the secrets of specific origination.

It is desirable, then, to see what facts there are in animal organization which point to innate conditions (powers and tendencies), as yet unexplained, and upon which the theory of "Natural Selection" is unable to throw any explanatory light.

The facts to be considered are the phenomena of "homology," and especially of serial, bilateral, and vertical homology.

The word "homology" indicates such a relation between two parts that they may be said in some sense to be "the same," or at least "of similar nature." This similarity, however, does not relate to the *use* to which parts are put, but only to their relative position with regard to other parts, or to their

mode of origin. There are many kinds of homology,[1] but it is
only necessary to consider the three kinds above enumerated.

The term "homologous" may be applied to parts in two
individual animals of different kinds, or to different parts of
the same individual. Thus "the right and left hands," or
"joints of the backbone," or "the teeth of the two jaws," are
homologous parts of the same individual. But the arm of a
man, the fore-leg of the horse, the paddle of the whale, and the
wing of the bat and the bird are all also homologous parts, yet

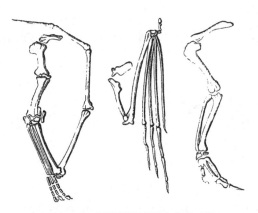

'WINGBONES OF PTERODACTYLE, BAT, AND BIRD.

of another kind, *i.e.* they are the same parts existing in animals
of different species.

On the other hand, the wing of the humming-bird and the
wing of the humming-bird moth are not homologous at all, or
in any sense; for the resemblance between them consists solely
in the use to which they are put, and is therefore only a relation
of *analogy.* There is no relation of *homology* between them,
because they have no common resemblance as to their relations
to surrounding parts, or as to their mode of origin. Similarly,

---

[1] For an enumeration of the more obvious homological relationships
see Ann. and Mag. of Nat. Hist. for August 1870, p. 118.

there is no homology between the wing of the bat and that of the flying-dragon, for the latter is formed of certain ribs, and not of limb bones.

SKELETON OF THE FLYING DRAGON.
(Showing the elongated ribs which support the flitting organ.)

Homology may be further distinguished into (1) a relationship which, on evolutionary principles, would be due to descent from a common ancestor, as the homological relation between the arm-bone of the horse and that of the ox, or between the singular ankle bones of the two lemurine genera, cheirogaleus and galago, and which relation has been termed by Mr. Ray Lankester "homogeny;"[1] and (2) a relationship induced, not derived—such as exists between parts closely similar in relative position, but with no genetic affinity, or only a remote one, as the homological relation between the chambers of the heart of a

[1] See Ann. and Mag. of Nat. Hist., July 1870.

bat and those of a bird, or the similar teeth of the thylacine
and the dog before spoken of. For this relationship Mr. Ray
Lankester has proposed the term "homoplasy."

TARSAL BONES OF DIFFERENT LEMUROIDS.
(Right tarsus of Galago; left tarsus of Cheirogaleus.)

"Serial homology" is a relation of resemblance existing between two or more parts placed in series one behind the other in
the same individual. Examples of such homologues are the ribs,

A CENTIPEDE.

or joints of the backbone of a horse, or the limbs of a centipede.
The latter animal is a striking example of serial homology.
The body (except at its two ends) consists of a longitudinal
series of similar segments. Each segment supports a pair
of limbs, and the appendages of all the segments (except as
before) are completely alike.

A less complete case of serial homology is presented by Crustacea (animals of the crab class), notably by the squilla and
by the common lobster. In the latter animal we have a six-

SQUILLA.

jointed abdomen (the so-called tail), in front of which is a
large solid mass (the cephalo-thorax), terminated anteriorly by
a jointed process (the rostrum). On the under surface of the
body we find a quantity of moveable appendages. Such are, *e.g.*,

PART OF THE SKELETON OF THE LOBSTER.

feelers (Fig. 9), jaws (Figs. 6, 7, and 8), foot-jaws (Fig. 5), claws
and legs (Figs. 3 and 4), beneath the cephalo-thorax; and flat
processes (Fig. 2), called "swimmerets," beneath the so-called
tail or abdomen.

Now, these various appendages are distinct and different

enough as we see them in the adult, but they all appear
in the embryo as buds of similar form and size, and the

thoracic limbs at first consist each of two members, as the swimmerets always do.

This shows what great differences may exist in
size, in form, and in function, between parts which
are developmentally the same, for all these appendages are modifications of one common kind of
structure, which becomes differently modified in
different situations; in other words, they are
serial homologues.

The segments of the body, as they follow one
behind the other, are also serially alike, as is
plainly seen in the abdomen or tail. In the
cephalo-thorax of the lobster, however, this is
disguised. It is therefore very interesting to
find that in the other crustacean before mentioned, the squilla, the segmentation of the body
is more completely preserved, and even the first
three segments, which go to compose the head,
remain permanently distinct.

Such an obvious and unmistakeable serial repetition of parts does not obtain in the highest,
or backboned animals, the Vertebrata. Thus in
man and other mammals, nothing of the kind is
externally visible, and we have to penetrate to his
skeleton to find such a series of homologous parts.

There, indeed, we discover a number of pairs
of bones, each pair so obviously resembling the
others, that they all receive a common name—the

SPINE OF GALAGO
ALLENII.

ribs. There also (*i.e.* in the skeleton) we find a
still more remarkable series of similar parts, the
joints of the spine or backbone (vertebræ), which are admitted
by all to possess a certain community of structure.

It is in their limbs, however, that the Vertebrata present the most obvious and striking serial homology—almost the only serial homology noticeable externally.

The facts of serial homology seem hardly to have excited the amount of interest they certainly merit.

Very many writers, indeed, have occupied themselves with investigations and speculations as to what portions of the leg and foot answer to what parts of the arm and hand, a question which has only recently received a more or less satisfactory solution through the successive concordant efforts of Professor Humphry,[1] Professor Huxley,[2] the Author of this work,[3] and Professor Flower.[4] Very few writers, however, have devoted much time or thought to the question of serial homology in general. Mr. Herbert Spencer, indeed, in his very interesting "First Principles of Biology," has given forth ideas on this subject, which are well worthy careful perusal and consideration, and some of which apply also to the other kinds of homology mentioned above. He would explain the serial homologies of such creatures as the lobster and centipede thus: Animals of a very low grade propagate themselves by spontaneous fission. If certain creatures found benefit from this process of division remaining incomplete, such creatures (on the theory of "Natural Selection") would transmit their selected tendency to such incomplete division to their posterity. In this way, it is conceivable, that animals might arise in the form of long chains of similar segments, each of which chains would consist of a number of imperfectly separated individuals, and be equivalent to a series of separate individuals belonging to kinds in which the fission was complete. In other words, Mr.

---

[1] Treatise on the Human Skeleton, 1858.
[2] Hunterian Lectures for 1864.
[3] Linnæan Transactions, vol. xxv. p. 395, 1866.
[4] Hunterian Lectures for 1870, and Journal of Anat. for May 1870.

Spencer would explain it as the coalescence of organisms of a lower degree of aggregation in one longitudinal series, through survival of the fittest aggregations. This may be so. It is certainly an ingenious speculation, but facts have not yet been brought forward which demonstrate it. Had they been so, this kind of serial homology might be termed "homogenetic."

The other kind of serial repetitions, namely, those of the vertebral column, are explained by Mr. Spencer as the results of alternate strains and compressions acting on a primitively homogeneous cylinder. The serial homology of the fore and hind limbs is explained by the same writer as the result of a similarity in the influences and conditions to which they are exposed. Serial homologues so formed might be called, as Mr. Ray Lankester has proposed, "homoplastic." But there are, it is here contended, abundant reasons for thinking that the predominant agent in the production of the homologies of the limbs is an *internal* force or tendency. And if such a power can be shown to be necessary in this instance, it may also be legitimately used to explain such serial homologies as those of the centipede's segments and of the joints of the backbone. At the same time it is not, of course, pretended that external conditions do not contribute their own effects in addition. The presence of this internal power will be rendered more probable if valid arguments can be brought forward against the explanations which Mr. Herbert Spencer has offered.

Lateral homology (or bilateral symmetry) is the resemblance between the right and left sides of an animal, or of part of an animal ; as, *e.g.*, between our right hand and our left. It exists more or less at one or other time of life in all animals, except some very lowly organized creatures. In the highest animals this symmetry is laid down at the very dawn of life, the first trace of the future creature being a longitudinal streak—the embryonic "primitive groove." This kind of homology is explained by Mr. Spencer as the result of the similar way in which conditions affect the right and left sides respectively.

Vertical homology (or vertical symmetry) is the resemblance existing between parts which are placed one above the other beneath. It is much less general and marked than serial, or lateral homology. Nevertheless, it is plainly to be seen in the tail region of most fishes, and in the far-extending dorsal (back) and ventral (belly) fins of such kinds as the sole and the flounder.

It is also strikingly shown in the bones of the tail of certain efts, as in *Chioglossa*, where the complexity of the upper (neural) arch is closely repeated by the inferior one. Again, in *Spelerpes rubra*, where almost vertically ascending articular processes above are repeated by almost vertically descending articular processes below. Also in the axolotl, where there are douple pits, placed side by side, not only superiorly but at the same time inferiorly.[1]

This kind of homology is also explained by Mr. Spencer as the result of the similarity of conditions affecting the two parts. Thus he explains the very general absence of symmetry between the dorsal and ventral surfaces of animals by the different conditions to which these two surfaces are respectively exposed, and in the same way he explains the asymmetry of the flat-fishes (*Pleuronectidæ*), of snails, &c.

VERTEBRÆ
OF AXOLOTL.

Now, first, as regards Mr. Spencer's explanation of animal forms by means of the influence of external conditions, the following observations may be made. Abundant instances are brought forward by him of admirable adaptation of structure to circumstances, but as to the immense majority of these it is very difficult, if not impossible, to see *how* external

---

[1] See a Paper on the "Axial Skeleton of the Urodela," in Proc. Zool. Soc. 1870, p. 266.

conditions can have produced, or even tended to have produced them. For example, we may take the migration of one eye of the sole to the other side of its head. What is there here either in the darkness, or the friction, or in any other conceivable

PLEURONECTIDÆ, WITH THE PECULIARLY PLACED EYE IN DIFFERENT POSITIONS.

external cause, to have produced the first beginning of such an unprecedented displacement of the eye? Mr. Spencer has beautifully illustrated that correlation which all must admit to exist between the forms of organisms and their surrounding external conditions, but by no means proved that the latter are *the cause* of the former.

Some internal conditions (or in ordinary language some internal power and force) must be conceded to living organisms, otherwise incident forces must act upon them and upon non-living aggregations of matter in the same way and with similar effects.

If the mere presence of these incident forces produces so ready a response in animals and plants, it must be that there are, in their case, conditions disposing and enabling them so to respond, according to the old maxim, *Quicquid recipitur, recipitur ad modum recipientis,* as the same rays of light which bleach a piece of silk, blacken nitrate of silver. If, therefore, we

attribute the forms of organisms to the action of external conditions, *i.e.* of incident forces on their modifiable structure, we
give but a partial account of the matter, removing a step back,
as it were, the action of the internal condition, power, or force
which must be conceived as occasioning such ready modifiability.
But indeed it is not at all easy to see how the influence of the
surface of the ground or any conceivable condition or force can
produce the difference which exists between the ventral and
dorsal shields of the carapace of a tortoise, or by what differences
of merely external causes the ovaries of the two sides of the
body can be made equal in a bat and unequal in a bird.

There is, on the other hand, an *a priori* reason why we should
expect to find that the symmetrical forms of all animals are due
to internal causes. This reason is the fact that the symmetrical
forms of minerals are undoubtedly due to such causes. It is

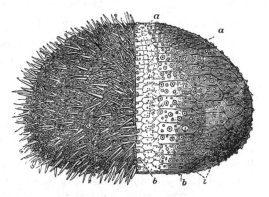

AN ECHINUS, OR SEA-URCHIN.
(The spines removed from one-half.)

unnecessary here to do more than allude to the beautiful and
complex forms presented by inorganic structures. With regard to
organisms, however, the wonderful Acanthometræ and the Poly-
cystina may be mentioned as presenting complexities of form
which can hardly be thought to be due to other than *internal*
causes. The same may be said of the great group of Echino-

derms, with their amazing variety of component parts. If then internal forces can so build up the most varied structures, they are surely capable of producing the serial, lateral, and vertical symmetries which higher animal forms exhibit. Mr. Spencer is the more bound to admit this, inasmuch as in his doctrine of "physiological units" he maintains that these organic atoms of his have an innate power of building up and evolving the whole and perfect animal from which they were in each case derived. To build up and evolve the various symmetries here spoken of is not one whit more mysterious. Directly to refute Mr. Spencer's assertion, however, would require the bringing forward of examples of organisms which are ill-adapted to their positions, and out of harmony with their surroundings —a difficult task indeed.[1]

Secondly, as regards the last-mentioned author's explanation of such serial homology as exists in the centipede and its allies, the very groundwork is open to objection. Multiplication by spontaneous fission seems from some recent researches to be much less frequent than has been supposed, and more evidence is required as to the fact of the habitual propagation of *any* planariæ in this fashion.[2] But even if this were as asserted,

[1] Just as Buffon's superfluous lament over the unfortunate organization of the sloth has been shown, by the increase of our knowledge, to have been uncalled for and absurd, so other supposed instances of non-adaptation will, no doubt, similarly disappear. Mr. Darwin, in his "Origin of Species," 5th edition, p. 220, speaks of a woodpecker (*Colaptes campestris*) as having an organization quite at variance with its habits, and as never climbing a tree, though possessed of the special arboreal structure of other woodpeckers. It now appears, however, from the observations of Mr. W. H. Hudson, C.M.Z.S., that its habits are in harmony with its structure. See Mr. Hudson's third letter to the Zoological Society, published in the Proceedings of that Society for March 24, 1870, p. 159.

[2] Dr. Cobbold has informed the Author that he has never observed a planaria divide spontaneously, and he is sceptical as to that process taking place at all. Dr. H. Charlton Bastian has also stated that, in spite of much observation, he has never seen the process in *vorticella*.

nevertheless it fails to explain the peculiar condition presented by *Syllis* and some other annelids, where a new head is formed at intervals in certain segments of the body.   Here there is evidently an innate tendency to the development at intervals of a complex whole.   It is not the budding out or spontaneous

AN ANNELID DIVIDING SPONTANEOUSLY.
(A new head having been formed towards the hinder end of the body of the parent.)

fission of certain segments, but the transformation in a definite and very peculiar manner of parts which already exist into other and more complex parts.   Again, the processes of development presented by some of these creatures do not by any means point to an origin through the linear coalescence of primitively distinct animals by means of imperfect segmentation. Thus in certain Diptera (two winged flies) the legs, wings, eyes, &c., are derived from masses of formative tissue (termed imaginal disks), which by their mutual approximation together build up parts of the head and body,[1] recalling to mind the development of Echinoderms.

Again, Nicholas Wagner found in certain other Diptera, the Hessian flies, that the larva gives rise to secondary larvæ within it, which develop and burst the body of the primary larva. The secondary larvæ give rise, similarly, to another set within them, and these again to another[2] set.

Again, the fact that in *Tænia echinococcus* one egg produces numerous individuals, tends to invalidate the argument that the increase of segments during development is a relic of specific genesis.

Mr. H. Spencer seems to deny serial homology to the mollusca, but it is difficult to see why the shell segments of chiton are not such homologues because the segmentation is superficial. Similarly the external processes of eolis, doris, &c., are good examples of serial homology, as also are plainly the successive chambers of the orthoceratidæ. Nor are parts of a series less serial, because arranged spirally, as in most gasteropods. Mr. Spencer observes of the molluscous as of the vertebrate animal, "You cannot cut it into transverse slices, each of which contains a digestive organ, a respiratory organ, a reproductive organ, &c."[3] But the same may be said of every single arthropod and annelid if it be meant that all these organs are not contained in every possible slice. While if it be meant that parts of all such organs are contained in certain slices, then some of the mollusca may also be included.

Another objection to Mr. Spencer's speculation is derived from considerations which have already been stated, as to past

[1] Professor Huxley's Hunterian Lecture, March 16, 1868.
[2] Ibid. March 18.   .   [3] "Principles of Biology," vol. ii. p. 105.

time. For if the annulose animals have been formed by aggregation, we ought to find this process much less perfect in the oldest form. But a complete development, such as already obtains in the lobster, &c., was reached by the Eurypterida and Trilobites of the palæozoic strata; and annelids, probably formed

TRILOBITE.

mainly like those of the present day, abounded during the deposition of the oldest fossiliferous rocks.

Thirdly, and lastly, as regards such serial homology as is exemplified by the backbone of man, there are also several objections to Mr. Spencer's mechanical explanation.

On the theory of evolution most in favour, the first Vertebrata were aquatic. Now, as natation is generally effected by repeated and vigorous lateral flexions of the body, we ought to find the segmentation much more complete laterally than on the dorsal and ventral aspects of the spinal column. Nevertheless, in those species which, taken together, constitute a series of more and more distinctly segmented forms, the segmentation gradually increases *all round* the central part of the spinal column.

Mr. Spencer[1] thinks it probable that the sturgeon has retained the notochordal (that is, the primitive, unsegmented) structure

[1] "Principles of Biology," vol. ii. p. 203.

because it is sluggish.   But Dr. Günther informs me that the
sluggishness of the common tope (*Galeus vulgaris*) is much
like that of the sturgeon, and yet the bodies of its vertebræ
are distinct and well-ossified.   Moreover, the great salamander
of Japan is much more inert and sluggish than either, and yet
it has a well-developed, bony spine.

I can learn nothing of the habits of the sharks *Hexanchus*,
Heptanchus, and *Echinorhinus*, but Müller describes them as
possessing a persistent *chorda dorsalis*).[1]   It may be they have
the habits of the tope, but other sharks are amongst the very
swiftest and most active of fishes.

In the bony pike (*lepidosteus*), the rigidity of the bony scales
by which it is completely enclosed must prevent any excessive
flexion of the body, and yet its vertebral column presents a
degree of ossification and vertebral completeness greater than
that found in any other fish whatever.

Mr. Spencer supports his argument by the non-segmentation
of the anterior end of the skeletal axis, *i.e.* by the non-segmentation of the skull.   But in fact the skull *is* segmented, and,
according to the quasi-vertebral theory of the skull put forward by Professor Huxley,[2] is probably formed of a number of
coalesced segments, of some of which the trabeculæ cranii and
the mandibular and hyoidean arches are indications.   What is,
perhaps, most remarkable however is, that the segmentation of
the skull—its separation into the three occipital, parietal, and
frontal elements—is most complete and distinct in the highest
class, and this can have nothing, however remotely, to do with
the cause suggested by Mr. Spencer.

Thus, then, there is something to be said in opposition to
both the aggregational and the mechanical explanations of serial
homology.   The explanations suggested are very ingenious, yet

[1] Quoted by H. Stannius in his "Handbuch der Anatomie der Wirbel-
thiere," Zweite Auflage, Erstes Buch, § 7, p. 17.
[2] In his last Hunterian Course of Lectures, 1869.

repose upon a very small basis of fact. Not but that the process of vertebral segmentation may have been sometimes assisted by the mechanical action suggested.

It remains now to consider what are the evidences in support of the existence of an internal power, by the action of which these homological manifestations are evolved. It is here contended that there *is* good evidence of the existence of some such special internal power, and that not only from facts of comparative anatomy, but also from those of teratology [1] and pathology. These facts appear to show, not only that there are homological internal relations, but that they are so strong and energetic as to re-assert and re-exhibit themselves in creatures which, on the Darwinian theory, are the descendants of others in which they were much less marked. They are, in fact, sometimes even more plain and distinct in animals of the highest types than in inferior forms, and, moreover, this deep-seated tendency acts even in diseased and abnormal conditions.

Mr. Darwin recognizes [2] these homological relations, and does " not doubt that they may be mastered more or less completely by Natural Selection." He does not, however, give any explanation of these phenomena other than the imposition on them of the name " laws of correlation ;" and indeed he says, " The nature of the bond of correlation is frequently quite obscure." Now, it is surely more desirable to make use, if possible, of one conception than to imagine a number of, to all appearance, separate and independent " laws of correlation " between different parts of each animal.

But even some of these alleged laws hardly appear well founded. Thus Mr. Darwin, in support of such a law of concomitant variation as regards hair and teeth, brings forward the

---

[1] " The Science of Abnormal Forms."

[2] " Animals and Plants under Domestication," vol. ii. p. 322 ; and " Origin of Species," 5th edition, 1869, p. 178.

case of Julia Pastrana,[1] and a man of the Burmese Court, and
adds,[2] "These cases and those of the hairless dogs forcibly call
to mind the fact that the two orders of mammals, namely, the
Edentata and Cetacea, which are the most abnormal in their
dermal covering, are likewise the most abnormal either by
deficiency or redundancy of teeth." The assertion with regard
to these orders is certainly true, but it should be borne in mind
at the same time that the armadillos, which are much more abnormal than are the American anteaters as regards their dermal
covering, in their dentition are less so. The Cape ant-eater, on
the other hand, the Aard-vark (Orycteropus), has teeth formed

THE AARD-VARK (ORYCTEROPUS).

on a type quite different from that existing in any other mammal;
yet its hairy coat is not known to exhibit any such strange
peculiarity. Again, those remarkable scaly ant-eaters of the Old
World—the pangolins (Manis)—stand alone amongst mammals as

[1] A remarkable woman exhibited in London a few years ago.
[2] "Animals and Plants under Domestication," vol. ii. p. 328.

regards their dermal covering; having been classed with lizards by early naturalists on account of their clothing of scales, yet their mouth is like that of the hairy ant-eaters of the New World. On the other hand, the duck-billed platypus of Aus-

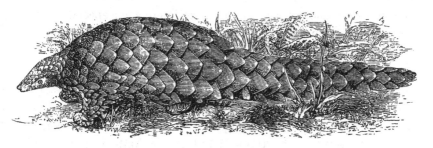

THE PANGOLIN (MANIS).

tralia (Ornithorhynchus) is the only mammal which has teeth formed of horn, yet its furry coat is normal and ordinary. Again, the Dugong and Manatee are dermally alike, yet extremely different as regards the structure and number of their teeth. The porcupine also, in spite of its enormous

DUGONG.

armature of quills, is furnished with as good a supply of teeth as are the hairy members of the same family, but not with a better one; and in spite of the deficiency of teeth in the hairless dogs, no converse redundancy of teeth has, it is believed, been remarked in Angora cats and rabbits. To say the least, then, this law of correlation presents numerous and remarkable exceptions.

To return, however, to the subject of homological relations: it is surely inconceivable that indefinite variation with survival of the fittest can ever have built up these serial, bilateral, and vertical homologies, without the action of some special innate power or tendency so to build up, possessed by the organism itself in each case. By "special tendency" is meant one the laws and conditions of which are as yet unknown, but which is analogous to the innate power and tendency possessed by crystals similarly, to build up certain peculiar and very definite forms.

First, with regard to comparative anatomy. The correspondence between the thoracic and pelvic limbs is notorious. Professor Gegenbaur has lately endeavoured [1] to explain this resemblance by the derivation of each limb from a primitive form of fin. This fin is supposed to have had a marginal external (radial) series of cartilages, each of which supported a series of secondary cartilages, starting from the inner (ulnar) side of the distal part of the supporting marginal piece. The root marginal piece would become the humerus or femur, as the case might be : the second marginal piece, with the piece attached to the inner side of the distal end of the root marginal piece, would together form either the radius and ulna or the tibia and fibula, and so on.

Now there is little doubt (from *a priori* considerations) but that the special differentiation of the limb bones of the higher Vertebrates has been evolved from anterior conditions existing in some fish-like form or other. But the particular view advocated by the learned Professor is open to criticism. Thus, it may be objected against this view, first, that it takes no account of the radial ossicle which becomes so enormous in the mole ; secondly, that it does not explain the extra series of

---

[1] " Ueber das Gliedmaassenskelet der Enaliosaurier, Jenaischen Zeitschrift," Bd. v. Heft 3, Taf. xiii.

ossicles which are formed on the *outer* (radial or marginal) side of
the paddle in the Ichthyosaurus ; and thirdly, and most importantly, that even if this had been the way in which the limbs
had been differentiated, it would not be at all inconsistent
with the possession of an innate power of producing, and an
innate tendency to produce similar and symmetrical homological
resemblances.　It would not be so because resemblances of the
kind are found to exist, which, on the Darwinian theory,
must be subsequent and secondary, not primitive and ancestral.
Thus we find in animals of the eft kind (certain amphibians),

SKELETON OF AN ICHTHYOSAURUS.

in which the tarsus is cartilaginous, that the carpus is cartilaginous likewise.　And we shall see in cases of disease and
of malformation what a tendency there is to a similar affection of homologous parts.　In efts, as Professor Gegenbaur
himself has pointed out,[1] there is a striking correspondence
between the bones or cartilages supporting the arm, wrist,
and fingers, and those sustaining the leg, ankle, and toes,
with the exception that the toes exceed the fingers in number
by one.

　　Yet these animals are far from being the root-forms from which
all the Vertebrata have diverged, as is evidenced from the degree
of specialization which their structure presents.　If they have

---

[1] In his work on the Carpus and Tarsus.

descended from such primitive forms as Professor Gegenbaur
imagines, then they have built up a secondary serial homology—
a repetition of similar modifications—fully as remarkable as if it

A. SKELETON OF ANTERIOR EXTREMITY OF AN EFT.
B. SKELETON OF POSTERIOR EXTREMITY OF THE SAME.

were primary. The Plesiosauria—those extinct marine reptiles
of the Secondary period, with long necks, small heads, and paddle-
like limbs—are of yet higher organization than are the efts and

SKELETON OF A PLESIOSAURUS.

other Amphibia. Nevertheless they present us with a similarity
of structure between the fore and hind limb, which is so great

as almost to be identity. But the Amphibia and Plesiosauria, though not themselves primitive vertebrate-types, may be thought by some to have derived their limb-structure by direct descent from such. Tortoises, however, must be admitted to be not only highly differentiated organisms, but to be far indeed removed from primeval vertebrate structure. Yet certain tortoises[1] (notably *Chelydra Temminckii*) exhibit such a remarkable uniformity in fore and hind limb structure (extending even up to the proximal ends of the humerus and femur) that it is impossible to doubt its independent development in these forms.

Again in the Potto (Perodicticus) there is an extra bone in the foot, situated in the transverse ligament enclosing the flexor tendons. It is noteworthy that in the *hand* of the same animal a serially homologous structure should also be developed.[2] In the allied form called the slow lemur (Nycticebus) we have certain arrangements of the muscles and tendons of the hand which reproduce in great measure those of the foot and *vice versâ*.[3] And in the Hyrax another myological resemblance appears.[4] It is, however, needless to multiply instances which can easily be produced in large numbers if required.

Secondly, with regard to teratology, it is notorious that similar abnormalities are often found to co-exist in both the pelvic and thoracic limbs.

M. Isidore Geoffroy St. Hilaire remarks,[5] "L'anomalie se répète d'un membre thoracique au membre abdominal du même côté." And he afterwards quotes from Weitbrecht,[6] who had "observé dans un cas l'absence simultanée aux deux mains

---

[1] An excellent specimen displaying this resemblance is preserved in the Museum of the Royal College of Surgeons.

[2] Phil. Trans. 1867, p. 353.

[3] Proc. Zool. Soc. 1865, p. 255.

[4] Ibid. p. 351.

[5] "Hist. Générale des Anomalies," t. i. p. 228. Bruxelles, 1837.

[6] Nov. Comment. Petrop. t. ix. p. 269.

et aux deux pieds, de quelques doigts, de quelques metacarpiens et metatarsiens, enfin de quelques os du carpe et du tarse."

LONG FLEXOR MUSCLES AND TENDONS OF THE HAND.

P.t. Pronator teres. *F.s.* Flexor sublimis digitorum. *F.p.* Flexor profundus digitorum. *F l.p.* Flexor longus pollicis.

Professor Burt G. Wilder, in his paper on extra digits,[1] has

[1] Read on June 2, 1868, before the Massachusetts Medical Society. See vol. ii. No. 3.

recorded no less than twenty-four cases where such excess co-existed in both little fingers ; also one case in which the right little finger and little toe were so affected; six in which it was both the little fingers and both the little toes; and twenty-two other cases more or less the same, but in which the details were not accurately to be obtained.

Mr. Darwin cites[1] a remarkable instance of what he is inclined to regard as the development in the foot of birds of a sort of representation of the wing-feathers of the hand. He says: " In several distinct breeds of the pigeon and fowl the legs and the two outer toes are heavily feathered, so that, in the trumpeter pigeon, they appear like little wings. In the feather-legged bantam, the 'boots,' or feathers, which grow from the outside of the leg, and generally from the two outer toes, have, according to the excellent authority of Mr. Hewitt, been seen to exceed the wing-feathers in length, and in one case were actually nine and a half inches in length! As Mr. Blyth has remarked to me, these leg-feathers resemble the primary wing-feathers, and are totally unlike the fine down which naturally grows on the legs of some birds, such as grouse and owls. Hence it may be suspected that excess of food has first given redundancy to the plumage, and then that the law of homologous variation has led to the development of feathers on the legs, in a position corresponding with those on the wing, namely, on the outside of the tarsi and toes. I am strengthened in this belief by the following curious case of correlation, which for a long time seemed to me utterly inexplicable,—namely, that in pigeons of any breed, if the legs are feathered, the two outer toes are partially connected by skin. These two outer toes correspond with our third and fourth toes. Now, in the wing of the pigeon, or any other bird, the first and fifth digits are wholly aborted; the second is rudimentary, and carries the so-called ' bastard wing ;' whilst the third and fourth

[1] " Animals and Plants under Domestication," vol. ii. p. 322.

digits are completely united and enclosed by skin, together forming the extremity of the wing. So that in feather-footed pigeons not only does the exterior surface support a row of long feathers like wing-feathers, but the very same digits which in the wing are completely united by skin become partially united by skin in the feet; and thus, by the law of the correlated variation of homologous parts, we can understand the curious connexion of feathered legs and membrane between the outer toes."

Irregularities in the circulating system are far from uncommon, and sometimes illustrate this homological tendency. My friend and colleague Mr. George G. Gascoyen, assistant surgeon at St. Mary's Hospital, has supplied me with two instances of symmetrical affections which have come under his observation.

In the first of these the brachial artery bifurcated almost at its origin, the two halves re-uniting at the elbow-joint, and then dividing into the radial and ulnar arteries in the usual manner. In the second case an aberrant artery was given off from the radial side of the brachial artery, again almost at its origin. This aberrant artery anastomosed below the elbow-joint with the radial side of the radial artery. In each of these cases the right and left sides varied in precisely the same manner.

Thirdly, as to pathology. Mr. James Paget,[1] speaking of symmetrical diseases, says: "A certain morbid change of structure on one side of the body is repeated in the exactly corresponding part of the other side." He then quotes and figures a diseased lion's pelvis from the College of Surgeons Museum, and says of it: "Multiform as the pattern is in which the new bone, the product of some disease comparable with a human rheumatism, is deposited—a pattern more complex and irregular than the spots upon a map—there is not one spot or line on one side which is not represented, as exactly as it would be in a mirror, on the other. The likeness has more than

[1] "Lectures on Surgical Pathology," 1853, vol. i. p. 18.

daguerreotype exactness." He goes on to observe : "I need not describe many examples of such diseases. Any out-patients' room will furnish abundant instances of exact symmetry in the eruptions of eczema, lepra, and psoriasis; in the deformities of chronic rheumatism, the paralyses from lead; in the eruptions excited by iodide of potassium or copaiba. And any large museum will contain examples of equal symmetry in syphilitic ulcerations of the skull; in rheumatic and syphilitic deposits on the tibiæ and other bones; in all the effects of chronic rheumatic arthritis, whether in the bones, the ligaments, or the cartilages; in the fatty and earthy deposits in the coats of arteries."[1]

He also considered it to be proved that, " Next to the parts which are symmetrically placed, none are so nearly identical in composition as those which are homologous. For example, the backs of the hands and of the feet, or the palms and soles, are often not only symmetrically, but similarly, affected with psoriasis. So are the elbows and the knees; and similar portions of the thighs and the arms may be found affected with ichthyosis. Sometimes also specimens of fatty and earthy deposits in the arteries occur, in which exact similarity is shown in the plan, though not in the degree, with which the disease affects severally the humeral and femoral, the radial and peroneal, the ulnar and posterior tibial arteries."

Dr. William Budd[2] gives numerous instances of symmetry in disease, both lateral and serial. Thus, amongst others, we have one case (William Godfrey), in which the hands and feet were distorted. "The distortion of the right hand is greater than that of the left, of the right foot greater than that of the left foot." In another (Elizabeth Alford) lepra affected the extensor surfaces of the thoracic and pelvic limbs. Again, in the case of skin disease illustrated in Plate III., "The analogy between

1 " Lectures on Surgical Pathology," 1853, vol. i. p. 22.
2 See "Medico-Chirurgical Transactions," vol. xxv. (or vii. of 2nd series), 1842, p. 100, Pl. III.

the elbows and knees is clearly expressed in the fact that these were the only parts affected with the disease." [1]

Professor Burt Wilder,[2] in his paper on "Pathological Polarities," strongly supports the philosophical importance of these peculiar relations, adding arguments in favour of antero-posterior homologies, which it is here unnecessary to discuss, enough having been said, it is believed, to thoroughly demonstrate the existence of these deep internal relations which are named lateral and serial homologies.

What explanation can be offered of these phenomena? To say that they exhibit a " nutritional relation " brought about by a " balancing of forces " is merely to give a new denomination to the unexplained fact. The changes are, *of course,* brought about by a " nutritional " process, and the symmetry is undoubtedly the result of a "balance of forces," but to say so is a truism. The question is, what is the cause of this " nutritional balancing " ? It is here contended that it must be due to an internal cause which at present science is utterly incompetent to explain. It is an internal property possessed by each living organic whole as well as by each non-living crystalline mass, and that there is such internal power or tendency, which may be spoken of as a " polarity, " seems to be demonstrated by the instances above given, which can easily be multiplied indefinitely. Mr. Herbert Spencer [3] (speaking of the reproduction, by budding, of a Begonia-leaf) recognizes a power of the kind. He says, " We have, therefore, no alternative but to say that the living particles composing one of these fragments have an innate tendency to arrange themselves into the shape of the organism to which they belong. We must infer that a plant or animal of any species is made up of special units, in all of which there dwells the intrinsic apti-

[1] Med.-Chirurg. Trans. vol. xxv. (or vii. of 2nd series), 1842, p. 122.
[2] See *Boston Medical and Surgical Journal* for April 5, 1866, vol. lxxiv. p. 189.
[3] " Principles of Biology," vol. i. p. 180.

tude to aggregate into the form of that species ; just as in the
atoms of a salt, there dwells the intrinsic aptitude to crystallize
in a particular way.   It seems difficult to conceive that this can
be so ; but we see that it *is* so." . . . . "For this property
there is no fit term.   If we accept the word polarity as a name
for the force by which inorganic units are aggregated into a
form peculiar to them, we may apply this word to the analogous
force displayed by organic limits."

Dr. Jeffries Wyman,[1] in his paper on the "Symmetry and
Homology of Limbs," has a distinct chapter on the "Analogy
between Symmetry and Polarity," illustrating it by the effects
of magnets on "particles in a polar condition."

Mr. J. J. Murphy, after noticing[2] the power which crystals
have to repair injuries inflicted on them and the modifications
they undergo through the influence of the medium in which
they may be formed, goes on to say :[3] "It needs no proof that
in the case of spheres and crystals the forms and the structures
are the effect, and not the cause, of the formative principles.
Attraction, whether gravitative or capillary, produces the
spherical form ; the spherical form does not produce attraction.
And crystalline polarities produce crystalline structure and
form ; crystalline structure and form do not produce crystalline
polarities.   The same is not quite so evident of organic forms,
but it is equally true of them also." . . . . . "It is not conceivable that the microscope should reveal peculiarities of
structure corresponding to peculiarities of habitual tendency in
the embryo, which at its first formation has no structure whatever ;"[4] and he adds that "there is something quite inscrutable and mysterious" in the formation of a new individual

[1] See the "Proceedings of the Boston Society of Natural History,"
vol. xi. June 5, 1867.

[2] "Habit and Intelligence," vol. i. p. 75.

[3] Ibid. p. 112.   [4] Ibid. p. 170.

from the germinal matter of the embryo. In another place[1] he says: "We know that in crystals, notwithstanding the variability of form within the limits of the same species, there are definite and very peculiar formative laws, which cannot possibly depend on anything like organic functions, because crystals have no such functions; and it ought not to surprise us if there are similar formative or morphological laws among organisms, which, like the formative laws of crystallization, cannot be referred to any relation of form or structure to function. Especially, I think, is this true of the lowest organisms, many of which show great beauty of form, of a kind that appears to be altogether due to symmetry of growth; as the beautiful star-like rayed forms of the *acanthometræ*, which are low animal organisms not very different from the Foraminifera." Their "definiteness of form does not appear to be accompanied by any corresponding differentiation of function between different parts; and, so far as I can see, the beautiful regularity and symmetry of their radiated forms are altogether due to unknown laws of symmetry of growth, just like the equally beautiful and somewhat similar forms of the compound six-rayed, star-shaped crystals of snow."

Altogether, then, it appears that each organism has an innate tendency to develop in a symmetrical manner, and that this tendency is controlled and subordinated by the action of external conditions, and not that this symmetry is superinduced only *ab externo*. In fact, that each organism has its own internal and special laws of growth and development.

If, then, it is still necessary to conceive an internal law or "substantial form," moulding each organic being,[2] and directing

---

[1] "Habit and Intelligence," vol. i. p. 229.

[2] It is hardly necessary to say that the Author does not mean that there is, in addition to a real objective crystal, another real, objective separate thing beside it, namely the "force" directing it. All that is meant is that the action of the crystal in crystallizing must be *ideally* separated from the crystal itself, not that it is *really* separate.

its development as a crystal is built up, only in an indefinitely more complex manner, it is congruous to imagine the existence of some internal law accounting at the same time for specific divergence as well as for specific identity.

A principle regulating the successive evolution of different organic forms is not one whit more mysterious than is the mysterious power by which a particle of structureless sarcode develops successively into an egg, a grub, a chrysalis, a butterfly, when all the conditions, cosmical, physical, chemical, and vital, are supplied, which are the requisite accompaniments to determine such evolution.

CHAPTER IX.

EVOLUTION AND ETHICS.

The origin of morals an inquiry not foreign to the subject of this book.—
Modern utilitarian view as to that origin.—Mr. Darwin's speculation
as to the origin of the abhorrence of incest.—Cause assigned by him
insufficient.—Care of the aged and infirm opposed by "Natural Selection;" also self-abnegation and asceticism.—Distinctness of the ideas
"right" and "useful."—Mr. John Stuart Mill.—Insufficiency of
"Natural Selection" to account for the origin of the distinction
between duty and profit.—Distinction of moral acts into "material"
and "formal."—No ground for believing that formal morality exists
in brutes.—Evidence that it does exist in savages.—Facility with
which savages may be misunderstood.—Objections as to diversity of
customs.—Mr. Hutton's review of Mr. Herbert Spencer.—Anticipatory
character of morals.—Sir John Lubbock's explanation.—Summary
and conclusion.

Any inquiry into the origin of the notion of "morality"—the
conception of "right"—may, perhaps, be considered as somewhat
remote from the question of the Genesis of Species; the more
so, since Mr. Darwin, at one time, disclaimed any pretension to
explain the origin of the higher psychical phenomena of man.
His disciples, however, were never equally reticent, and indeed
he himself is now not only about to produce a work on
man (in which this question must be considered), but he has
distinctly announced the extension of the application of his
theory to the very phenomena in question. He says:[1] "In the

---

[1] "Origin of Species," 5th edition, 1839, p. 577.

distant future I see open fields for far more important researches. Psychology will be based on a new foundation, that of the necessary acquirement of each mental power and capacity by gradation. Light will be thrown on the origin of man and his history." It may not be amiss then to glance slightly at the question, so much disputed, concerning the origin of ethical conceptions and its bearing on the theory of "Natural Selection."

The followers of Mr. John Stuart Mill, of Mr. Herbert Spencer, and apparently, also, of Mr. Darwin, assert that in spite of the great *present* difference between the ideas "useful" and "right," yet that they are, nevertheless, one in *origin*, and that that origin consisted ultimately of pleasurable and painful sensations.

They say that "Natural Selection" has evolved moral conceptions from perceptions of what was useful, *i.e.* pleasurable, by having through long ages preserved a predominating number of those individuals who have had a natural and spontaneous liking for practices and habits of mind useful to the race, and that the same power has destroyed a predominating number of those individuals who possessed a marked tendency to contrary practices. The descendants of individuals so preserved have, they say, come to inherit such a liking and such useful habits of mind, and that at last (finding this inherited tendency thus existing in themselves, distinct from their tendency to conscious self-gratification) they have become apt to regard it as fundamentally distinct, *innate*, and independent of all experience. In fact, according to this school, the idea of "right" is only the result of the gradual accretion of useful predilections which, from time to time, arose in a series of ancestors naturally selected. In this way, "morality" is, as it were, the congealed past experience of the race, and "virtue" becomes no more than a sort of "retrieving," which the thus improved human animal practises by a perfected and inherited habit, regardless of self-gratification, just as the brute animal has acquired the habit of seeking prey and bringing it to his master, instead of devouring it himself.

Though Mr. Darwin has not as yet expressly advocated this view, yet some remarks made by him appear to show his disposition to sympathise with it. Thus, in his work on "Animals and Plants under Domestication," [1] he asserts that "the savages of Australia and South America hold the crime of incest in abhorrence;" but he considers that this abhorrence has probably arisen by "Natural Selection," the ill effects of close interbreeding causing the less numerous and less healthy offspring of incestuous unions to disappear by degrees, in favour of the descendants (greater both in number and strength) of individuals who naturally, from some cause or other, as he suggests, preferred to mate with strangers rather than with close blood-relations ; this preference being transmitted and becoming thus instinctive, or habitual, in remote descendants.

But on Mr. Darwin's own ground, it may be objected that this notion fails to account for "abhorrence," and "moral reprobation ;" for, as no stream can rise higher than its source, the original "slight feeling" which was *useful* would have been perpetuated, but would never have been augmented beyond the degree requisite to ensure this beneficial preference, and therefore would not certainly have become magnified into "abhorrence." It will not do to assume that the union of males and females, each possessing the required "slight feeling," must give rise to offspring with an intensified feeling of the same kind ; for, apart from reversion, Mr. Darwin has called attention to the unexpected modifications which sometimes result from the union of *similarly* constituted parents. Thus, for example, he tells us : [2] "If two top-knotted canaries are matched, the young, instead of having very fine top-knots, are generally bald." From examples of this kind, it is fair, on Darwinian principles, to infer that the

[1] Vol. ii. p. 122.
[2] "Animals and Plants under Domestication," vol. i. p. 295.

union of parents who possessed a similar inherited aversion might
result in phenomena quite other than the augmentation of such
aversion, even if the two aversions should be altogether similar;
while, very probably, they might be so different in their nature
as to tend to neutralize each other. Besides, the union of parents
so similarly emotional would be rare indeed amongst savages,
where marriages would be owing to almost anything rather than to
congeniality of mind between the spouses. Mr. Wallace tells us,[1]
that they choose their wives for "rude health and physical beauty,"
and this is just what might be naturally supposed. Again, we
must bear in mind the necessity there is that *many individuals*
should be similarly and simultaneously affected with this aversion from consanguineous unions; as we have seen in the second
chapter, how infallibly variations presented by only a few individuals, tend to be eliminated by mere force of numbers. Mr.
Darwin indeed would throw back this aversion, if possible, to a
pre-human period; since he speculates as to whether the gorillas
or orang-utans, in effecting their matrimonial relations, show any
tendency to respect the prohibited degrees of affinity.[2] No tittle
of evidence, however, has yet been adduced pointing in any such
direction, though surely if it were of such importance and efficiency as to result (through the aid of " Natural Selection "
alone) in that "abhorrence" before spoken of, we might expect
to be able to detect unmistakeable evidence of its incipient stages.
On the contrary, as regards the ordinary apes (for with regard to
the highest there is no evidence of the kind) as we see them
in confinement, it would be difficult to name any animals less
restricted, by even a generic bar, in the gratification of the
sexual instinct. And although the conditions under which they
have been observed are abnormal, yet these are hardly the
animals to present us in a state of nature, with an extraordinary and exceptional sensitiveness in such matters.

1 " Natural Selection," p. 350.
2 " Animals and Plants under Domestication," vol. ii.

To take an altogether different case. Care of, and tenderness towards, the aged and infirm are actions on all hands admitted to be "right;" but it is difficult to see how such actions could ever have been so useful to a community as to have been seized on and developed by the exclusive action of the law of the "survival of the fittest." On the contrary, it seems probable that on strict utilitarian principles the rigid political economy of Tierra del Fuego would have been eminently favoured and diffused by the impartial action of "Natural Selection" alone. By the rigid political economy referred to, is meant that destruction and utilization of "useless mouths" which Mr. Darwin himself describes in his highly interesting "Journal of Researches."[1] He says : "It is certainly true, that when pressed in winter by hunger, they kill and devour their old women before they kill their dogs. The boy being asked why they did this, answered, 'Doggies catch otters, old women no.' They often run away into the mountains, but they are pursued by the men and brought back to the slaughter-house at their own firesides." Mr. Edward Bartlett, who has recently returned from the Amazons, reports that at one Indian village where the cholera made its appearance, the whole population immediately dispersed into the woods, leaving the sick to perish uncared for and alone. Now, had the Indians remained, undoubtedly far more would have died ; as doubtless, in Tierra del Fuego, the destruction of the comparatively useless old women has often been the means of preserving the healthy and reproductive young. Such acts surely must be greatly favoured by the stern and unrelenting action of exclusive "Natural Selection."

In the same way that admiration which all feel for acts of self-denial done for the good of others, and tending even towards the destruction of the actor, could hardly be accounted for on Darwinian principles alone ; for self-immolators must but rarely leave direct descendants, while the community they benefit must

---

[1] See 2nd edition, vol. i. p. 214.

by their destruction tend, so far, to morally deteriorate. But
devotion to others of the same community is by no means
all that has to be accounted for.   Devotion to the whole
human race, and devotion to God—in the form of asceticism—
have been and are very generally recognized as "good;" and
the Author contends that it is simply impossible to conceive
that such ideas and sanctions should have been developed
by "Natural Selection" alone, from only that degree of un-
selfishness necessary for the preservation of brutally barbarous
communities in the struggle for life.   That degree of unselfish-
ness once attained, further improvement would be checked by
the mutual opposition of diverging moral tendencies and spontaneous variations in all directions.   Added to which, we have the
principle of reversion and atavism, tending powerfully to restore
and reproduce that more degraded anterior condition whence the
later and better state painfully emerged.

Very few, however, dispute the complete distinctness, here
and now, of the ideas of "duty" and "interest" whatever may
have been the origin of those ideas.   No one pretends that
ingratitude may, in any past abyss of time, have been a virtue,
or that it may be such now in Arcturus or the Pleiades.   Indeed,
a certain eminent writer of the utilitarian school of ethics has
amusingly and very instructively shown how radically distinct
even in his own mind are the two ideas which he nevertheless
endeavours to identify.   Mr. John Stuart Mill, in his examination of "Sir William Hamilton's Philosophy," says,[1] if "I am
informed that the world is ruled by a being whose attributes are
infinite, but what they are we cannot learn, nor what the principles
of his government, except that 'the highest human morality
which we are capable of conceiving' does not sanction them;
convince me of it, and I will bear my fate as I may.   But when
I am told that I must believe this, and at the same time call this

[1] Page 103.

O

being by the names which express and affirm the highest human morality, I say in plain terms that I will not. Whatever power such a being may have over me, there is one thing which he shall not do : he shall not compel me to worship him. I will call no being good, who is not what I mean when I apply that epithet to my fellow-creatures ; and if such a being can sentence me to hell for not so calling him, to hell I will go."

This is unquestionably an admirable sentiment on the part of Mr. Mill (with which every absolute moralist will agree), but it contains a complete refutation of his own position, and is a capital instance [1] of the vigorous life of moral intuition in one who professes to have eliminated any fundamental distinction between the "right" and the "expedient." For if an action is morally good, and to be done, merely in proportion to the amount of pleasure it secures, and morally bad and to be avoided as tending to misery, and if it could be *proved* that by calling God good—whether He is so or not, in our sense of the term,—we could secure a maximum of pleasure, and by refusing to do so we should incur endless torment, clearly, on utilitarian principles, the flattery would be good.

Mr. Mill, of course, must also mean that, in the matter in question, all men would do well to act with him. Therefore, he must mean that it would be well for all to accept (on the hypothesis above given) infinite and final misery for all as the result of the pursuit of happiness as the only end.

It must be recollected that in consenting to worship this unholy God, Mr. Mill is not asked to do harm to his neighbour, so that his refusal reposes simply on his perception of the immorality of the requisition. It is also noteworthy that an omnipotent Deity is supposed incapable of altering Mr. Mill's mind and moral perceptions.

Mr. Mill's decision is right, but it is difficult indeed to see

[1] I have not the merit of having noticed this inconsistency ; it was pointed out to me by my friend the Rev. W. W. Roberts.

how, without the recognition of an "absolute morality," he can justify so utter and final an abandonment of all utility in favour of a clear and distinct moral perception.

These two ideas, the "right" and the "useful," being so distinct here and now, a greater difficulty meets us with regard to their origin from some common source, than met us before when considering the first beginnings of certain bodily structures. For the distinction between the "right" and the "useful" is so fundamental and essential that not only does the idea of benefit not enter into the idea of duty, but we see that the very fact of an act *not* being beneficial to us makes it the more praiseworthy, while gain tends to diminish the merit of an action. Yet this idea, "right," thus excluding, as it does, all reference to utility or pleasure, has nevertheless to be constructed and evolved from utility and pleasure, and ultimately from pleasurable sensations, if we are to accept pure Darwinianism : if we are to accept, that is, the evolution of man's psychical nature and highest powers, by the exclusive action of "Natural Selection," from such faculties as are possessed by brutes ; in other words, if we are to believe that the conceptions of the highest human morality arose through minute and fortuitous variations of brutal desires and appetites in all conceivable directions.

It is here contended, on the other hand, that no conservation of any such variations could ever have given rise to the faintest beginning of any such moral perceptions ; that by "Natural Selection" alone the maxim *fiat justitia, ruat cœlum* could never have been excogitated, still less have found a widespread acceptance ; that it is impotent to suggest even an approach towards an explanation of the *first beginning* of the idea of "right." It need hardly be remarked that acts may be distinguished not only as pleasurable, useful, or beautiful, but also as good in two different senses : (1) *materially* moral acts, and (2) acts which are *formally* moral. The first are acts good in themselves, *as acts*, apart from any intention of the agent which may or may not have been directed towards "right." The second are acts which are good not only in themselves, as acts, but also in the deliberate *intention* of the agent who recognizes his actions as being "right." Thus acts may be *materially* moral or immoral, in a very high degree, without being in the least *formally* so. For example, a person may tend and minister to a sick man with scrupulous care and exactness, having in view all the time nothing but the future reception of a good legacy. Another may, in the dark, shoot his own father, taking him to be an assassin, and so commit what is *materially* an act of parricide, though *formally* it is only an act of self-defence of more or less culpable rashness. A woman may innocently, because ignorantly, marry a married man, and so commit a *material* act of adultery. She may discover the facts, and persist, and so make her act *formal* also.

Actions of brutes, such as those of the bee, the ant, or the beaver, however materially good as regards their relation to the community to which such animals belong, are absolutely destitute of the most incipient degree of real, *i.e.* formal "goodness," because unaccompanied by mental acts of conscious will directed towards the fulfilment of duty. Apology is due for thus stating so elementary a distinction, but the statement is not superfluous, for confusion of thought, resulting from confounding together these very distinct things, is unfortunately far from uncommon.

Thus some Darwinians assert that the germs of morality exist in brutes, and we have seen that Mr. Darwin himself speculates on the subject as regards the highest apes. It may safely be affirmed, however, that there is no trace in brutes of any actions simulating morality which are not explicable by the fear of punishment, by the hope of pleasure, or by personal affection. No sign of moral reprobation is given by any brute, and yet had such existed in germ through Darwinian abysses of past time, some evidence of its existence must surely have been rendered perceptible through "survival of the fittest" in other forms besides man, if that "survival" has alone and exclusively produced it in him.

Abundant examples may, indeed, be brought forward of useful acts which simulate morality, such as parental care of the young, &c. But did the most undeviating habits guide all brutes in such matters, were even aged and infirm members of a community of insects or birds carefully tended by young which benefited by their experience, such acts would not indicate even the faintest rudiment of real, *i.e.* formal, morality. "Natural Selection" would, of course, often lead to the prevalence of acts beneficial to a community, and to acts *materially* good; but unless they can be shown to be *formally* so, they are not in the least to the point, they do not offer any explanation of the origin of an altogether new and fundamentally different motive and conception.

It is interesting, on the other hand, to note Mr. Darwin's statement as to the existence of a distinct moral feeling, even in, perhaps, the very lowest and most degraded of all the human races known to us. Thus in the same "Journal of Researches"[1] before quoted, bearing witness to the existence of moral reprobation on the part of the Fuegians, he says: "The nearest approach to religious feeling which I heard of was shown by York Minster (a Fuegian so named), who, when Mr. Bynoe shot some very young ducklings as specimens, declared in the most solemn manner, 'Oh, Mr. Bynoe, much rain, snow, blow much.' This was evidently a retributive punishment for wasting human food."

Mr. Wallace gives the most interesting testimony, in his "Malay Archipelago," to the existence of a very distinct, and in some instances highly developed moral sense in the natives with whom he came in contact. In one case,[2] a Papuan who had been paid in advance for bird-skins and who had not been able

[1] Vol. i. p. 215. [2] "Malay Archipelago," vol. ii. p. 365

to fulfil his contract before Mr. Wallace was on the point of starting, "came running down after us holding up a bird, and saying with great satisfaction, ' Now I owe you nothing!'" And this though he could have withheld payment with complete impunity.

Mr. Wallace's observations and opinions on this head seem hardly to meet with due appreciation in Sir John Lubbock's recent work on Primitive Man.[1] But considering the acute powers of observation and the industry of Mr. Wallace, and especially considering the years he passed in familiar and uninterrupted intercourse with natives, his opinion and testimony should surely carry with it great weight. He has informed the Author that he found a strongly marked and widely diffused modesty, in sexual matters, amongst all the tribes with which he came in contact. In the same way Mr. Bonwick, in his work on the Tasmanians, testifies to the modesty exhibited by the naked females of that race, who by the decorum of their postures gave evidence of the possession in germ of what under circumstances would become the highest chastity and refinement.

Hasty and incomplete observations and inductions are prejudicial enough to physical science, but when their effect is to degrade untruthfully our common humanity, there is an additional motive to regret them. A hurried visit to a tribe, whose language, traditions and customs are unknown, is sometimes deemed sufficient for "smart" remarks as to "ape characters," &c., which are as untrue as irrelevant. It should not be forgotten how extremely difficult it is to enter into the ideas and feelings of an alien race. If in the nineteenth century a French theatrical audience can witness with acquiescent approval, as a type of English manners and ideas, the representation of a marquis who sells his wife at Smithfield, &c. &c., it is surely no

[1] "The Origin of Civilization and the Primitive Condition of Man," p. 261. Longmans, 1870.

wonder if the ideas of a tribe of newly visited savages should be more or less misunderstood. To enter into such ideas requires long and familiar intimacy, like that experienced by the explorer of the Malay Archipelago. From him, and others, we have abundant evidence that moral ideas exist, at least in germ, in savage races of men, while they sometimes attain even a highly developed state. No amount of evidence as to acts of moral depravity is to the point, as the object here aimed at is to establish that moral intuitions *exist* in savages, not that their actions are good.

Objections, however, are sometimes drawn from the different notions as to the moral value of certain acts, entertained by men of various countries or of different epochs ; also from the difficulty of knowing what particular actions in certain cases are the right ones, and from the effects which prejudice, interest, passion, habit, or even, indirectly, physical conditions, may have upon our moral perceptions. Thus Sir John Lubbock speaks [1] of certain Feejeeans, who, according to the testimony of Mr. Hunt,[2] have the custom of piously choking their parents under certain circumstances, in order to insure their happiness in a future life. Should any one take such facts as telling *against* the belief in an absolute morality, he would show a complete misapprehension of the point in dispute ; for such facts tell in *favour* of it.

Were it asserted that man possesses a distinct innate power and faculty by which he is made intuitively aware what acts considered in and by themselves are right and what wrong,—an infallible and universal internal code,—the illustration would be to the point. But all that need be contended for is that the intellect perceives not only truth, but also a quality of " higher " which ought to be followed, and of " lower " which ought to be avoided ; when two lines of conduct are presented to the will for choice, the intellect so acting being the conscience.

[1] " Primitive Man," p. 248.
[2] " Fiji and the Fijians," vol. i. p. 183.

This has been well put by Mr. James Martineau in his excellent essay on Whewell's Morality. He says,[1] "If moral good were a quality resident in each action, as whiteness in snow, or sweetness in fruits; and if the moral faculty was our appointed instrument for detecting its presence; many consequences would ensue which are at variance with fact. The wide range of differences observable in the ethical judgments of men would not exist; and even if they did, could no more be reduced and modified by discussion than constitutional differences of hearing or of vision. And, as the quality of moral good either must or must not exist in every important operation of the will, we should discern its presence or absence separately in each; and even though we never had the conception of more than one insulated action, we should be able to pronounce upon its character. This, however, we have plainly no power to do. Every moral judgment is relative, and involves a comparison of two terms. When we praise what *has been* done, it is with the coexistent conception of something *else* that *might have been* done; and when we resolve on a course as right, it is to the exclusion of some other that is wrong. This fact, that every ethical decision is in truth a *preference*, an election of one act as higher than another, appears of fundamental importance in the analysis of the moral sentiments."

From this point of view it is plain how trifling are arguments drawn from the acts of a savage, since an action highly immoral in us might be one exceedingly virtuous in him—being the highest presented to his choice in his degraded intellectual condition and peculiar circumstances.

It need only be contended, then, that there *is* a perception of "right" incapable of further analysis; not that there is any infallible internal guide as to all the complex actions which

---

[1] "Essays," Second Series, vol. ii. p. 13.

present themselves for choice. The *principle* is given in our nature, the *application* of the principle is the result of a thousand educational influences.

It is no wonder, then, that, in complex "cases of conscience," it is sometimes a matter of exceeding difficulty to determine which of two courses of action is the less objectionable. This no more invalidates the truth of moral principles than does the difficulty of a mathematical problem cast doubt on mathematical principles. Habit, education, and intellectual gifts facilitate the correct application of both.

Again, if our moral insight is intensified or blunted by our habitual wishes or, indirectly, by our physical condition, the same may be said of our perception of the true relations of physical facts one to another. An eager wish for marriage has led many a man to exaggerate the powers of a limited income, and a fit of dyspepsia has given an unreasonably gloomy aspect to more than one balance-sheet.

Considering that moral intuitions have to do with *insensible* matters, they cannot be expected to be more clear than the perception of physical facts. And if the latter perceptions may be influenced by volition, desire, or health, our moral views may also be expected to be so influenced, and this in a higher degree because they so often run counter to our desires. A bottle or two of wine may make a sensible object appear double; what wonder, then, if our moral perceptions are sometimes warped and distorted by such powerful agencies as an evil education or an habitual absence of self-restraint. In neither case does occasional distortion invalidate the accuracy of normal and habitual perception.

The distinctness here and now of the ideas of "right" and "useful" is however, as before said, fully conceded by Mr. Herbert Spencer, although he contends that these conceptions are one in root and origin.

His utilitarian Genesis of Morals, however, has been recently combated by Mr. Richard Holt Hutton in a paper which appeared in *Macmillan's Magazine.*[1]

This writer aptly objects an *argumentum ad hominem*, applying to morals the same argument that has been applied in this work to our sense of musical harmony, and by Mr. Wallace to the vocal organs of man.

Mr. Herbert Spencer's notions on the subject are thus expressed by himself: " To make my position fully understood, it seems needful to add that, corresponding to the fundamental propositions of a developed moral science, there have been, and still are developing in the race certain fundamental moral intuitions; and that, though these moral intuitions are the result of accumulated experiences of utility gradually organized and inherited, they have come to be quite independent of conscious experience. Just in the same way that I believe the intuition of space possessed by any living individual to have arisen from organized and consolidated experiences of all antecedent individuals, who bequeathed to him their slowly developed nervous organizations; just as I believe that this intuition, requiring only to be made definite and complete by personal experiences, has practically become a form of thought quite independent of experience;—so do I believe that the experiences of utility, organized and consolidated through all past generations of the human race, have been producing corresponding nervous modifications which, by continued transmissions and accumulation, have become in us certain faculties of moral intuition, active emotions responding to right and wrong conduct, which have no apparent basis in the individual experiences of utility. I also hold that, just as the space intuition responds to the exact demonstrations of geometry, and has its rough conclusions interpreted and verified by them, so will moral intuitions respond to the demonstrations of moral science, and will have their rough conclusions interpreted and verified by them."

[1] See No. 117, July 1869, p. 272.

Against this view of Mr. Herbert Spencer, Mr. Hutton objects —"1. That even as regards Mr. Spencer's illustration from geometrical intuitions, his process would be totally inadequate, since you could not deduce the necessary space intuition of which he speaks from any possible accumulations of familiarity with space relations. . . . We cannot *inherit* more than our fathers *had:* no amount of experience of facts, however universal, can give rise to that particular characteristic of intuitions and *a priori* ideas, which compels us to deny the possibility that in any other world, however otherwise different, our experience (as to space relations) could be otherwise.

" 2. That the case of moral intuitions is very much stronger.

" 3. That if Mr. Spencer's theory accounts for anything, it accounts not for the deepening of a sense of utility and inutility into right and wrong, but for the drying up of the sense of utility and inutility into mere inherent tendencies, which would exercise over us not *more* authority but *less*, than a rational sense of utilitarian issues.

" 4. That Mr. Spencer's theory could not account for the intuitional sacredness now attached to *individual* moral rules and principles, without accounting *a fortiori* for the general claim of the greatest happiness principle over us as the final moral intuition—which is conspicuously contrary to the fact, as not even the utilitarians themselves plead any instinctive or intuitive sanction for their great principle.

" 5. That there is no trace of positive evidence of any single instance of the transformation of a utilitarian rule of right into an intuition, since we find no utilitarian principle of the most ancient times which is now an accepted moral intuition, nor any moral intuition, however sacred, which has not been promulgated thousands of years ago, and which has not constantly had to stop the tide of utilitarian *objections* to its authority —and this age after age, in our own day quite as much as in days gone by. . . . . Surely, if anything is remarkable in the history of morality, it is the *anticipatory* character, if I may use the expression, of moral principles—the intensity and absoluteness with which they are laid down ages before the world has approximated to the ideal thus asserted."

Sir John Lubbock, in his work on Primitive Man before referred to, abandons Mr. Spencer's explanation of the genesis of morals while referring to Mr. Hutton's criticisms on the subject. Sir John proposes to substitute " deference to authority" instead of " sense of interest " as the origin of our conception of " duty," saying that what has been found to be beneficial has been traditionally inculcated on the young, and thus has become to be dissociated from " interest " in the mind, though the inculcation itself originally sprung from that source. This, however, when analysed, turns out to be a distinction without a difference. It is nothing but utilitarianism, pure and simple, after all. For it can never be intended that authority is obeyed because of an intuition that it *should be deferred to*, for that would be to admit the very principle of absolute morality which Sir John combats. It must be meant, then, that authority is obeyed through fear of the consequences of disobedience, or through pleasure felt in obeying the authority which commands. In the latter case we have "pleasure" as the end and no rudiment of the conception "duty." In the former we have fear of punishment, which appeals directly to the sense of "utility to the individual," and no amount of such a sense will produce the least germ of "ought" which is a conception different *in kind*, and in which the notion of "punishment" has no place. Thus, Sir John Lubbock's explanation only concerns a *mode* in which the sense of " duty " may be stimulated or appealed to, and makes no approximation to an explanation of its origin.

Could the views of Mr. Herbert Spencer, of Mr. Mill, or of Mr. Darwin on this subject be maintained, or should they come to be generally accepted, the consequences would be disastrous indeed! Were it really the case that virtue was a *mere kind of*

"*retrieving,*" then certainly we should have to view with apprehension the spread of intellectual cultivation, which would lead
the human "retrievers" to regard from a new point of view their
fetching and carrying. We should be logically compelled to
acquiesce in the vociferations of some continental utilitarians,
who would banish altogether the senseless words " duty " and
" merit ; " and then, one important influence which has aided
human progress being withdrawn, we should be reduced to hope
that in this case the maxim *cessante causa cessat ipse effectus*
might through some incalculable accident fail to apply.

It is true that Mr. Spencer tries to erect a safeguard
against such moral disruption, by asserting that for every immoral act, word, or thought, each man during this life receives
minute and exact retribution, and that thus a regard for individual self-interest will effectually prevent any moral catastrophe.
But by what means will he enforce the acceptance of a dogma
which is not only incapable of proof, but is opposed to the commonly received opinion of mankind in all ages? Ancient litera-
ture, sacred and profane, teems with protests against the successful evil-doer, and certainly, as Mr. Hutton observes,[1] " Honesty
must have been associated by our ancestors with many unhappy
as well as many happy consequences, and we know that in
ancient Greece dishonesty was openly and actually associated
with happy consequences. . . . . When the concentrated experience of previous generations was held, *not* indeed to justify,
but to excuse by utilitarian considerations, craft, dissimulation,
sensuality, selfishness."

This dogma is opposed to the moral consciousness of many as
to the events of their own lives ; and the Author, for one, believes
that it is absolutely contrary to fact.

History affords multitudes of instances, but an example may
be selected from one of the most critical periods of modern

[1] *Macmillan's Magazine,* No. 117, July 1869.

times. Let it be granted that Lewis the Sixteenth of France
and his queen had all the defects attributed to them by the
most hostile of serious historians; let all the excuses possible
be made for his predecessor, Lewis the Fifteenth, and also for
Madame de Pompadour, can it be pretended that there are
grounds for affirming that the vices of the two former so far
exceeded those of the latter, that their respective fates were
plainly and evidently just? that while the two former died in
their beds, after a life of the most extreme luxury, the others
merited to stand forth through coming time as examples of the
most appalling and calamitous tragedy?

This theme, however, is too foreign to the immediate matter
in hand to be further pursued, tempting as it is. But a passing
protest against a superstitious and deluding dogma may stand,—
a dogma which may, like any other dogma, be vehemently
asserted and maintained, but which is remarkable for being
destitute, at one and the same time, of both authoritative sanction and the support of reason and observation.

To return to the bearing of moral conceptions on "Natural
Selection," it seems that, from the reasons given in this chapter,
we may safely affirm—1. That "Natural Selection" could not have
produced, from the sensations of pleasure and pain experienced
by brutes, a higher degree of morality than was useful; therefore it could have produced any amount of "beneficial habits,"
but not abhorrence of certain acts as impure and sinful.

2. That it could not have developed that high esteem for
acts of care and tenderness to the aged and infirm which actually
exists, but would rather have perpetuated certain low social
conditions which obtain in some savage localities.

3. That it could not have evolved from ape sensations the
noble virtue of a Marcus Aurelius, or the loving but manly
devotion of a St. Lewis.

4. That, alone, it could not have given rise to the maxim
fiat justitia, ruat cœlum.

5. That the interval between material and formal morality is one altogether beyond its power to traverse.

Also, that the anticipatory character of moral principles is a fatal bar to that explanation of their origin which is offered to us by Mr. Herbert Spencer. And, finally, that the solution of that origin proposed recently by Sir John Lubbock is a mere version of simple utilitarianism, appealing to the pleasure or safety of the individual, and therefore utterly incapable of solving the riddle it attacks.

Such appearing to be the case as to the power of "Natural Selection," we, nevertheless, find moral conceptions—*formally* moral ideas—not only spread over the civilized world, but manifesting themselves unmistakeably (in however rudimentary a condition, and however misapplied) amongst the lowest and most degraded of savages. If from amongst these, individuals can be brought forward who seem to be destitute of any moral conception, similar cases also may easily be found in highly civilized communities. Such cases tell no more against moral intuitions than do cases of colour-blindness or idiotism tell against sight and reason. We have thus a most important and conspicuous fact, the existence of which is fatal to the theory of "Natural Selection," as put forward of late by Mr. Darwin and his most ardent followers. It must be remarked, however, that whatever force this fact may have against a belief in the origination of man from brutes by minute, fortuitous variations, it has no force whatever against the conception of the orderly evolution and successive manifestation of specific forms by ordinary natural law—even if we include amongst such the upright frame, the ready hand and massive brain of man himself.

CHAPTER X.

PANGENESIS.

A provisional hypothesis supplementing "Natural Selection."—Statement
of the hypothesis.—Difficulty as to multitude of gemmules.—As to
certain modes of reproduction.—As to formations without the requisite
gemmules.—Mr. Lewes and Professor Delpino.—Difficulty as to developmental force of gemmules.—As to their spontaneous fission.—
Pangenesis and Vitalism.—Paradoxical reality.—Pangenesis scarcely
superior to anterior hypotheses.—Buffon.—Owen.—Herbert
Spencer.—"Gemmules" as mysterious as "physiological units."—
Conclusion.

In addition to the theory of "Natural Selection," by which it
has been attempted to account for the origin of species, Mr.
Darwin has also put forward what he modestly terms "a provisional hypothesis" (that of *Pangenesis*), by which to account for
the origin of each and every individual form.

Now, though the hypothesis of Pangenesis is no necessary
part of "Natural Selection," still any treatise on specific
origination would be incomplete if it did not take into consideration this last speculation of Mr. Darwin. The hypothesis in
question may be stated as follows : That each living organism is
ultimately made up of an almost infinite number of minute particles, or organic atoms, termed "gemmules," each of which has
the power of reproducing its kind. Moreover, that these particles circulate freely about the organism which is made up of them,
and are derived from all the parts of all the organs of the less

remote ancestors of each such organism during all the states and stages of such several ancestors' existence ; and therefore of the several states of each of such ancestors' organs. That such a complete collection of gemmules is aggregated in each ovum and spermatozoon in most animals, and in each part capable of reproducing by gemmation (budding) in the lowest animals and in plants. Therefore in many of such lower organisms such a congeries of ancestral gemmules must exist in every part of their bodies, since in them every part is capable of reproducing by gemmation. Mr. Darwin must evidently admit this, since he says : " It has often been said by naturalists that each cell of a plant has the actual or potential capacity of reproducing the whole plant; but it has this power only in virtue of containing gemmules *derived from every part*." [1]

Moreover, these gemmules are supposed to tend to aggregate themselves, and to reproduce in certain definite relations to other gemmules. Thus, when the foot of an eft is cut off, its reproduction is explained by Mr. Darwin as resulting from the aggregation of those floating gemmules which come next in order to those of the cut surface, and the successive aggregations of the other kinds of gemmules which come after in regular order. Also, the most ordinary processes of repair are similarly accounted for, and the successive development of similar parts and organs in creatures in which such complex evolutions occur is explained in the same way, by the independent action of separate gemmules.

In order that each living creature may be thus furnished, the number of such gemmules in each must be inconceivably great Mr. Darwin says : [2] "In a highly organized and complex anima the gemmules thrown off from each different cell or unit throughout the body must be inconceivably numerous and minute. Each unit of each part, as it changes during development—and we

---

[1] " Animals and Plants under Domestication," vol. ii. p. 403.
[2] Ibid. p. 366.

P

know that some insects undergo at least twenty metamorphoses
—must throw off its gemmules. All organic beings, moreover,
include many dormant gemmules derived from their grandparents
and more remote progenitors, but not from all their progenitors.
These *almost infinitely numerous* and minute gemmules must be
included in each bud, ovule, spermatozoon, and pollen grain."
We have seen also that in certain cases a similar multitude of
gemmules must be included also in every considerable part of
the whole body of each organism, but where are we to stop?
There must be gemmules not only from every organ, but from
every component part of such organ, from every subdivision of
such component part, and from every cell, thread, or fibre entering into the composition of such subdivision. Moreover, not
only from all these, but from each and every successive stage
of the evolution and development of such successively more
and more elementary parts. At the first glance this new atomic
theory has charms from its apparent simplicity, but the attempt
thus to follow it out into its ultimate limits and extreme consequences seems to indicate that it is at once insufficient and
cumbrous.

Mr. Darwin himself is, of course, fully aware that there must
be *some* limit to this aggregation of gemmules. He says:[1]
"Excessively minute and numerous as they are believed to be,
an infinite number derived, during a long course of modification
and descent, from each cell of each progenitor, could not be
supported and nourished by the organism."

But apart from these matters, which will be more fully considered further on, the hypothesis not only does not appear to
account for certain phenomena which, in order to be a valid
theory, it ought to account for; but it seems absolutely to
conflict with patent and notorious facts.

How, for example, does it explain the peculiar reproduction

---

[1] "Animals and Plants under Domestication," vol. ii. p. 402.

which is found to take place in certain marine worms—certain annelids ?

In such creatures we see that, from time to time, one of the segments of the body gradually becomes modified till it assumes the condition of a head, and this remarkable phenomenon is

AN ANNELID DIVIDING SPONTANEOUSLY.
(A new head having been formed towards the hinder end of the body of the parent.)

repeated again and again, the body of the worm thus multiplying serially into new individuals which successively detach themselves from the older portion. The development of such a mode of reproduction by " Natural Selection " seems not less inexplicable than does its continued performance through the aid of " pangenesis." For how can gemmules attach themselves to others to which they do not normally or generally succeed? Scarcely less difficult to understand is the process of the stomach-carrying-off mode of metamorphosis before spoken of as existing in the Echinoderms. Next, as to certain patent and notorious facts: On the hypothesis of pangenesis, no creature can develop an organ unless it possesses the component gemmules which serve for its formation. No creature can possess such gemmules unless it inherits them from its parents, grandparents, or its less remote ancestors. Now, the Jews are remarkably scrupulous as to marriage, and rarely contract such a union with individuals not of their own race. This practice has gone on for thousands of years, and similarly also for thousands of years the rite of circumcision has been unfailingly and carefully performed. If then the hypothesis of pangenesis is well founded, that rite ought to be now absolutely or nearly superfluous from the necessarily continuous absence of certain gemmules through so many centuries and so many generations. Yet it is not at all so, and this fact seems to amount almost to an experimental demonstration that the hypothesis of pangenesis is an insufficient explanation of individual evolution.

Two exceedingly good criticisms of Mr. Darwin's hypothesis have appeared. One of these is by Mr. G. H. Lewes,[1] the other by Professor Delpino of Florence.[2] The latter gentleman gives a report of an observation made by him upon a certain plant, which observation adds force to what has just been said about the Jewish race. He says:[3] "If we examine and compare the numerous species of the genus *Salvia*, commencing with *Salvia officinalis*, which may pass as the main state of the genus, and

[1] See *Fortnightly Review*, New Series, vol. iii. April 1868, p. 352.
[2] This appeared in the *Rivista Contemporanea Nazionale Italiana*, and was translated and given to the English public in *Scientific Opinion* for September 29, October 6, and October 13, 1869, pp. 365, 391, and 407.
[3] See *Scientific Opinion*, of October 13, 1869, p. 407.

concluding with *Salvia verticillata*, which may be taken as the most highly developed form, and as the most distant from the type, we observe a singular phenomenon. The lower cell of each of the two fertile anthers, which is much reduced and different from the superior even in *Salvia officinalis*, is transmuted in other *salviæ* into an organ (nectarotheca) having a very different form and function, and finally disappears entirely in *Salvia verticillata*.

" Now, on one occasion, in a flower belonging to an individual of *Salvia verticillata*, and only on the left stamen, I observed a perfectly developed and pollinigerous lower cell, perfectly homologous with that which is normally developed in *Salvia officinalis*. This case of atavism is truly singular. According to the theory of Pangenesis, it is necessary to assume that all the gemmules of this anomalous formation, and therefore the mother-gemmule of the cell, and the daughter-gemmules of the special epidermic tissue, and of the very singular subjacent tissue of the endothecium, have been perpetuated, and transmitted from parent to offspring in a dormant state, and through a number of generations, such as startles the imagination, and leads it to refuse its consent to the theory of Pangenesis, however seductive it may be." This seems a strong confirmation of what has been here advanced.

The main objection raised against Mr. Darwin's hypothesis is that it (Pangenesis) requires so many subordinate hypotheses for its support, and that some of these are not tenable.

Professor Delpino considers [1] that as many as eight of these subordinate hypotheses are required, namely, that—

" 1. The emission of the gemmules takes place, or may take place in all states of the cell.

" 2. The quantity of gemmules emitted from every cell is very great.

" 3. The minuteness of the gemmules is extreme.

[1] See *Scientific Opinion* of September 29, 1869, p. 3C6.

" 4. The gemmules possess two sorts of affinity, one of which might be called *propagative*, and the other *germinative* affinity.

" 5. By means of the propagative affinity all the gemmules emitted by all the cells of the individual flow together and become condensed in the cells which compose the sexual organs, whether male or female (embryonal vesicle, cells of the embryo, pollen grains, fovilla, antherozoids, spermatozoids), and likewise flow together and become condensed in the cells which constitute the organs of a sexual or agamic reproduction (buds, spores, bulbilli, portions of the body separated by scission, &c.).

" 6. By means of the germinative affinity, every gemmule (except in cases of anomalies or monstrosities) can be developed only in cells homologous with the mother-cells of the cell from which they originated. In other words, the gemmules from any cell can only be developed in unison with the cell preceding it in due order of succession, and whilst in a nascent state.

" 7. Of each kind of gemmule a great number perishes; a great number remains in a dormant state through many generations in the bodies of descendants; the remainder germinate and reproduce the mother-cell.

" 8. Every gemmule may multiply itself by a process of scission into any number of equivalent gemmules."

Mr. Darwin has published a short notice in reply to Professor Delpino, in *Scientific Opinion* of October 20, 1869, p. 426. In this reply he admits the justice of Professor Delpino's attack, but objects to the alleged necessity of the first subordinate hypothesis, namely, that the emission of gemmules takes place in all states of the cell." But if this is not the case, then a great part of the utility and distinction of pangenesis is destroyed, or as Mr. Lewes justly says,[1] " If gemmules produce whole cells, we have the very power which was pronounced mysterious in larger organisms."

[1] *Fortnightly Review*, New Series, vol. iii. April 1868, p. 508.

Mr. Darwin also does not see the force of the objection to the power of self-division which must be asserted of the gemmules themselves if Pangenesis be true. The objection, however, appears to many to be formidable. To admit the power of spontaneous division and multiplication in such rudimentary structures, seems a complete contradiction. The gemmules, by the hypothesis of Pangenesis, are the ultimate organized components of the body, the absolute organic atoms of which each body is composed; how then *can* they be divisible? Any part of a gemmule would be an impossible (because a *less* than possible) quantity. If it is divisible into still smaller organic wholes, as a germ-cell is, it must be made up as the germ-cell is, of subordinate component atoms, which are then the *true* gemmules. This process may be repeated *ad infinitum*, unless we get to true organic atoms, the true gemmules, whatever they may be, and they necessarily will be incapable of any process of spontaneous fission. It is remarkable that Mr. Darwin brings forward in support of gemmule fission, the observation that " Thuret has seen the zoospore of an alga divide itself, and both halves germinate." Yet on the hypothesis of Pangenesis, the zoospore of an alga must contain gemmules from all the cells of the parent algæ, and from all the parts of all their less remote ancestors in all their stages of existence. What wonder then that such an excessively complex body should divide and multiply; and what parity is there between such a body and a gemmule? A steam-engine and a steel-filing might equally well be compared together.

Professor Delpino makes a further objection which, however, will only be of weight in the eyes of Vitalists. He says,[1] Pangenesis is not to be received because "it leads directly to the negation of a specific vital principle, co-ordinating and regulating all the movements, acts, and functions of the individuals in which it is incarnated. For Pangenesis of the individual is a

---

[1] *Scientific Opinion*, of October 13, 1869, p. 408.

term without meaning. If, in contemplating an animal of high
organization, we regard it purely as an aggregation of developed
gemmules, although these gemmules have been evolved successively one after the other, and one within the other, notwithstanding they elude the conception of the *real and true individual*, these problematical and invisible gemmules must be
regarded as so many individuals. Now, that real, true, living
individuals exist in nature, is a truth which is persistently
attested to us by our consciousness. But how, then, can we explain that a great quantity of dissimilar elements, like the atoms
of matter, can unite to form those perfect unities which we call
individuals, if we do not suppose the existence of a specific principle, proper to the individual but foreign to the component
atoms, which aggregates these said atoms, groups them into molecules, and then moulds the molecules into cells, the cells into
tissues, the tissues into organs, and the organs into apparatus?"

"But, it may be urged in opposition by the Pangenesists, your
vital principle is an unknown and irresolute *x*. This is true;
but, on the other hand, let us see whether Pangenesis produces
a clearer formula, and one free from unknown elements. The
existence of the gemmules is a first unknown element; the propagative affinity of the gemmules is a second; their germinative
affinity is a third; their multiplication by fission is a fourth—
and what an unknown element!"

"Thus, in Pangenesis, everything proceeds by force of unknown elements, and we may ask whether it is more logical to
prefer a system which assumes a multitude of unknown elements
to a system which assumes only a single one?"

Mr. Darwin appears, by "Natural Selection," to destroy the
reality of species, and by Pangenesis that of the individual.
Mr. Lewes observes[1] of the individual that "This whole is only
a subjective conception which summarizes the parts, and that in

---

[1] *Fortnightly Review*, New Series, vol. iii. April 1868, p. 509.

point of fact it is the parts which are reproduced." But the
parts are also, from the same point of view, merely subjective
until we come to the absolute organic atoms. These atoms, on
the other hand, are utterly invisible, intangible; indeed, in the
words of Mr. Darwin, inconceivable. Thus, then, it results from
the theories in question, that the organic world is reduced to
utter unreality as regards all that can be perceived by the
senses or distinctly imagined by the mind; while the only reality
consists of the invisible, the insensible, the inconceivable; in
other words, nothing is known that really is, and only the non-
existent can be known. A somewhat paradoxical outcome of
the speculations of those who profess to rely exclusively on the
testimony of sense. " *Les extrêmes se touchent,*" and extreme
sensationalism shakes hands with the " das seyn ist das nichts "
of Hegel.

Altogether the hypothesis of Pangenesis seems to be little, if
at all, superior to anterior hypotheses of a more or less similar
nature.

Apart from the atoms of Democritus, and apart also from the
speculations of mediæval writers, the molecules of Bonnet and of
Buffon almost anticipated the hypothesis of Pangenesis. According to the last-named author,[1] organic particles from every
part of the body assemble in the sexual secretions, and by their
union build up the embryo, each particle taking its due place,
and occupying in the offspring a similar position to that which it
occupied in the parents. In 1849 Professor Owen, in his treatise
on " Parthenogenesis," put forward another conception. According to this, the cells resulting from the subdivision of the
germ-cell preserve their developmental force, unless employed in
building up definite organic structures. In certain creatures, and
in certain parts of other creatures, germ-cells unused are stored

---

[1] " Histoire Naturelle, générale et particulière," tome ii. 1749, p. 327.
" Ces liqueurs séminales sont toutes deux un extrait de toutes les parties
du corps," &c.

up, and by their agency lost limbs and other mutilations are repaired. Such unused products of the germ-cell are also supposed to become located in the generative products.

According to Mr. Herbert Spencer, in his "Principles of Biology," each living organism consists of certain so-called "physiological units." Each of these units has an innate power and capacity, by which it tends to build up and reproduce the entire organism of which it forms a part, unless in the meantime its force is exhausted by its taking part in the production of some distinct and definite tissue—a condition somewhat similar to that conceived by Professor Owen.

Now, at first sight, Mr. Darwin's atomic theory appears to be more simple than any of the others. It has been objected that while Mr. Spencer's theory requires the assumption of an innate power and tendency in each physiological unit, Mr. Darwin's, on the other hand, requires nothing of the kind, but explains the evolution of each individual by purely mechanical conceptions. In fact, however, it is not so. Each gemmule, according to Mr. Darwin, is really the seat of powers, elective affinities, and special tendencies as marked and mysterious as those possessed by the physiological unit of Mr. Spencer, with the single exception that the former has no tendency to build up the whole living, complex organism of which it forms a part. Some may think this an important distinction, but it can hardly be so, for Mr. Darwin considers that his gemmule has the innate power and tendency to build up and transform itself into the whole living, complex cell of which it forms a part; and the one tendency is, in principle, fully as difficult to understand, fully as mysterious, as is the other. The difference is but one of degree, not of kind. Moreover, the one mystery in the case of the "physiological unit" explains all, while with regard to the gemmule, as we have seen, it has to be supplemented by other powers and tendencies, each distinct, and each in itself inexplicable and profoundly mysterious.

That there should be physiological units possessed of the power attributed to them, harmonizes with what has recently been put forward by Dr. H. Charlton Bastian; who maintains that under fit conditions the simplest organisms develop themselves into relatively large and complex ones. This is not supposed by him to be due to any inheritance of ancestral gemmules, but to direct growth and transformation of the most minute and the simplest organisms, which themselves, by all reason and analogy, owe their existence to immediate transformation from the inorganic world.

Thus, then, there are grave difficulties in the way of the reception of the hypothesis of Pangenesis, which moreover, if established, would leave the evolution of individual organisms, when thoroughly analysed, little if at all less mysterious or really explicable than it is at present.

As was said at the beginning of this chapter, "Pangenesis" and "Natural Selection" are quite separable and distinct hypotheses. The fall of one of these by no means necessarily includes that of the other. Nevertheless, Mr. Darwin has associated them closely together, and, therefore, the refutation of Pangenesis may render it advisable for those who have hitherto accepted "Natural Selection" to reconsider that theory.

CHAPTER XI.

SPECIFIC GENESIS.

Review of the statements and arguments of preceding chapters.—Cumulative argument against predominant action of "Natural Selection."—Whether anything positive as well as negative can be enunciated.—Constancy of laws of nature does not necessarily imply constancy of specific evolution.—Possible exceptional stability of existing epoch.—Probability that an internal cause of change exists.—Innate powers must be conceived as existing somewhere or other.—Symbolism of molecular action under vibrating impulses.—Professor Owen's statement.—Statement of the Author's view.—It avoids the difficulties which oppose "Natural Selection."—It harmonizes apparently conflicting conceptions.—Summary and conclusion.

HAVING now severally reviewed the principal biological facts which bear upon specific manifestation, it remains to sum up the results, and to endeavour to ascertain what, if anything, can be said *positively*, as well as negatively, on this deeply interesting question.

In the preceding chapters it has been contended, in the first place, that no mere survival of the fittest accidental and minute variations can account for the incipient stages of useful structures, such as, *e. g.*, the heads of flat-fishes, the baleen of whales, vertebrate limbs, the laryngeal structures of the new-born kangaroo, the pedicellariæ of Echinoderms, or for many of the facts of mimicry, and especially those last touches of mimetic perfection, where an insect not only mimics a leaf, but one worm-eaten and attacked by fungi.

Also, that structures like the hood of the cobra and the rattle of the rattlesnake seem to require another explanation.

Again, it has been contended that instances of colour, as in some apes; of beauty, as in some shell-fish; and of utility, as in many orchids, are examples of conditions which are quite beyond the power of Natural Selection to originate and develop.

Next, the peculiar mode of origin of the eye (by the simultaneous and concurrent modification of distinct parts), with the wonderful refinement of the human ear and voice, have been insisted on; as also, that the importance of all these facts is intensified through the necessity (admitted by Mr. Darwin) that many individuals should be similarly and simultaneously modified in order that slightly favourable variations may hold their own in the struggle for life, against the over-whelming force and influence of mere number.

Again, we have considered, in the third chapter, the great improbability that from minute variations in all directions alone and unaided, save by the survival of the fittest, closely similar structures should independently arise; though, on a non-Darwinian evolutionary hypothesis, their development might be expected *a priori.* We have seen, however, that there are many instances of wonderfully close similarity which are not due to genetic affinity; the most notable instance, perhaps, being that brought forward by Mr. Murphy, namely, the appearance of the same eye-structure in the vertebrate and molluscous sub-kingdoms. A curious resemblance, though less in degree, has also been seen to exist between the auditory organs of fishes and of Cephalopods. Remarkable similarities between certain placental and implacental mammals, between the bird's-head processes of Polyzoa and the pedicellariæ of Echinoderms, between Ichthyosauria and Cetacea, with very many other similar coincidences, have also been pointed out.

Evidence has also been brought forward to show that similarity is sometimes directly induced by very obscure conditions, at present quite inexplicable, *e. g.* by causes immediately connected with geographical distribution ; as in the loss of the tail in certain forms of Lepidoptera and in simultaneous modifications of colour in others, and in the direct modification of young English oysters, when transported to the shore of the Mediterranean.

Again, it has been asserted that certain groups of organic forms seem to have an innate tendency to remarkable developments of some particular kind, as beauty and singularity of plumage in the group of birds of paradise.

It has also been contended that there is something to be said in favour of sudden, as opposed to exceedingly minute and gradual, modifications, even if the latter are not fortuitous. Cases were brought forward, in Chapter IV., such as the bivalve just mentioned, twenty-seven kinds of American trees simultaneously and similarly modified, also the independent production of pony breeds, and the case of the English greyhounds in Mexico, the offspring of which produced directly acclimated progeny. Besides these, the case of the Normandy pigs, of *Datura tatula,* and also of the black-shouldered peacock, have been spoken of. The teeth of the labyrinthodon, the hand of the potto, the whalebone of whales, the wings of birds, the climbing tendrils of some plants, &c. have also been adduced as instances of structures, the origin and production of which are probably due rather to considerable modifications than to minute increments.

It has also been shown that certain forms which were once supposed to be especially transitional and intermediate (as, *e.g.,* the aye-aye) are really by no means so ; while the general rule, that the progress of forms has been "from the more general to the more special," has been shown to present remarkable exceptions, as, *e. g.*, Macrauchenia, the Glyptodon, and the sabre-toothed tiger (Machairodus).

Next, as to specific stability, it has been seen that there may be a certain limit to normal variability, and that if changes take place they may be expected *a priori* to be marked and considerable ones, from the facts of the inorganic world, and perhaps also of the lowest forms of the organic world. It has also been seen that with regard to minute spontaneous variations in races, there is a rapidly increasing difficulty in intensifying them, in any one direction, by ever such careful breeding. Moreover, it has appeared that different species show a tendency to variability in special directions, and probably in different degrees, and that at any rate Mr. Darwin himself concedes the existence of an internal barrier to change when he credits the goose with " a singularly inflexible organization ;" also, that he admits the presence of an *internal* proclivity to change when he speaks of " a whole organization seeming to have become plastic, and tending to depart from the parental type."

We have seen also that a marked tendency to reversion does exist, inasmuch as it sometimes takes place in a striking manner, as exemplified in the white silk fowl in England, *in spite of* careful selection in breeding.

Again, we have seen that a tendency exists in nature to eliminate hybrid races, by whatever means that elimination is effected, while no similar tendency bars the way to an indefinite blending of varieties. This has also been enforced by statements as to the prepotency of certain pollen of identical species, but of distinct races.

To all the preceding considerations have been added others derived from the relations of species to past time. It has been contended that we have as yet no evidence of minutely intermediate forms connecting uninterruptedly together undoubtedly distinct species. That while even " horse ancestry " fails to supply such a desideratum, in very strongly marked and exceptional kinds (such as the Ichthyosauria, Chelonia, and Anoura), the absence of links is very important and significant. For if every species, without exception, has arisen by minute modifications, it seems incredible that a small percentage of such transitional forms should not have been preserved. This, of course, is especially the case as regards the marine Ichthyosauria and Plesiosauria, of which such numbers of remains have been discovered.

Sir William Thomson's great authority has been seen to oppose itself to "Natural Selection," by limiting, on astronomical and physical grounds, the duration of life on this planet to about one hundred million years. This period, it has been contended, is not nearly enough on the one hand for the evolution of all organic forms by the exclusive action of mere minute, fortuitous variations; on the other hand, for the deposition of all the strata which must have been deposited, if minute fortuitous variation was the manner of successive specific manifestation.

Again, the geographical distribution of existing animals has been seen to present difficulties which, though not themselves insurmountable, yet have a certain weight when taken in conjunction with all the other objections.

The facts of homology, serial, bilateral and vertical, have also been passed in review. Such facts, it has been contended, are not explicable without admitting the action of what may most conveniently be spoken of as an *internal* power, the existence of which is supported by facts not only of comparative anatomy but of teratology and pathology also. "Natural Selection" also has been shown to be impotent to explain these phenomena, while the existence of such an internal power of homologous evolution diminishes the *a priori* improbability of an analogous law of specific origination.

All these various considerations have been supplemented by an endeavour to show the utter inadequacy of Mr. Darwin's theory with regard to the higher psychical phenomena of man (especially the evolution of moral conceptions), and with regard to the evolution of individual organisms by the action of

Pangenesis.   And it was implied that if Mr. Darwin's latter
hypothesis can be shown to be untenable, an antecedent doubt
is thus thrown upon his other conception, namely, the theory of
" Natural Selection."

A cumulative argument thus arises against the prevalent
action of " Natural Selection," which, to the mind of the Author,
is conclusive.   As before observed,   he was not originally
disposed to reject Mr. Darwin's fascinating theory.   Reiterate
endeavours to solve its difficulties have, however, had the
effect of convincing him that that theory as the one or as the
leading explanation of the successive evolution and manifestation of specific forms, is untenable.   At the same time he
admits fully that " Natural Selection" acts and must act, and
that it plays in the organic world a certain though a secondary
and subordinate part.

The one *modus operandi* yet suggested having been found
insufficient, the question arises, Can another be substituted
in its place ?   If not, can anything that is positive, and if anything, what, be said as to the question of specific origination ?

Now, in the first place, it is of course axiomatic that the
laws which conditioned the evolution of extinct and of existing
species are of as much efficacy at this moment as at any preceding period, that they *tend* to the manifestation of new forms as
much now as ever before.   It by no means necessarily follows,
however, that this tendency is actually being carried into
effect, and that new species of the higher animals and plants
are actually now produced.   They may be so or they may not,
according as existing circumstances favour, or conflict with,
the action of those laws.   It is possible that lowly organized
creatures may be continually evolved at the present day, the
requisite conditions being more or less easily supplied.   There
is, however, no similar evidence at present as to higher forms ;
while, as we have seen in Chapter VII., there are *a priori* considerations which militate against their being similarly evolved.

The presence of wild varieties and the difficulty which often exists in the determination of species are sometimes adduced as arguments that high forms are now in process of evolution. These facts, however, do not necessarily prove more than that some species possess a greater variability than others, and (what is indeed unquestionable) that species have often been unduly multiplied by geologists and botanists. It may be, for example, that Wagner was right, and that all the American monkeys of the genus cebus may be reduced to a single species or to two.

With regard to the lower organisms, and supposing views recently advanced to become fully established, there is no reason to think that the forms said to be evolved were new species, but rather reappearances of definite kinds which had appeared before and will appear again under the same conditions. In the same way, with higher forms similar conditions must educe similar results, but here practically similar conditions can rarely obtain because of the large part which "descent" and "inheritance" always play in such highly organized forms.

Still it is conceivable that different combinations at different times may have occasionally the same outcome just as the multiplications of different numbers may have severally the same result.

There are reasons, however, for thinking it possible that the human race is a witness of an exceptionally unchanging and stable condition of things, if the calculations of Mr. Croll are valid as to how far variations in the eccentricity in the earth's orbit together with the precession of the equinoxes have produced changes in climate. Mr. Wallace has pointed out [1] that the last 60,000 years having been exceptionally unchanging as

[1] See *Nature*, March 3, 1870, p. 454. Mr. Wallace says (referring to Mr. Croll's paper in the *Phil. May.*), "As we are now, and have been for 60,000 years, in a period of low eccentricity, *the rate of change of species during that time may be no measure of the rate that has generally obtained in past geological epochs.*"

regards these conditions, specific evolution may have been exceptionally rare. It becomes then possible to suppose that for a similar period stimuli to change in the manifestation of animal forms may have been exceptionally few and feeble,—that is, if the conditions of the earth's orbit have been as exceptional as stated. However, even if new species are actually now being evolved as actively as ever, or if they have been so quite recently, no conflict thence necessarily arises with the view here advocated. For it by no means follows that if some examples of new species have recently been suddenly produced from individuals of antecedent species, we ought to be able to put our fingers on such cases; as Mr. Murphy well observes [1] in a passage before quoted, "If a species were to come suddenly into being in the wild state, as the Ancon sheep did under domestication, how could we ascertain the fact? If the first of a newly-born species were found, the fact of its discovery would tell nothing about its origin. Naturalists would register it as a very rare species, having been only once met with, but they would have no means of knowing whether it were the first or the last of its race."

But are there any grounds for thinking that in the genesis of species an *internal* force or tendency interferes, co-operates with and controls the action of external conditions?

It is here contended that there are such grounds, and that though inheritance, reversion, atavism, Natural Selection, &c., play a part not unimportant, yet that such an internal power is a great, perhaps the main, determining agent.

It will, however, be replied that such an entity is no *vera causa;* that if the conception is accepted, it is no real explanation; and that it is merely a roundabout way of saying that the facts are as they are, while the cause remains unknown. To this it may be rejoined that for all who believe in the existence of the

---

[1] "Habit and Intelligence," vol. i. p. 344.

abstraction "force" at all, other than will, this conception of an internal force must be accepted and located somewhere— cannot be eliminated altogether; and that therefore it may as reasonably be accepted in this mode as in any other.

It was urged at the end of the third chapter that it is congruous to credit mineral species with an internal power or force. By such a power it may be conceived that crystals not only assume their external symmetry, but even repair it when injured. Ultimate chemical elements must also be conceived as possessing an innate tendency to form certain unions, and to cohere in stable aggregations. This was considered towards the end of Chapter VIII.

Turning to the organic world, even on the hypothesis of Mr. Herbert Spencer or that of Mr. Darwin, it is impossible to escape the conception of innate internal forces. With regard to the physiological units of the former, Mr. Spencer himself, as we have seen, distinctly attributes to them "an *innate* tendency" to evolve the parent form from which they sprang. With regard to the gemmules of Mr. Darwin, we have seen, in Chapter X., with how many innate powers, tendencies, and capabilities they must each be severally endowed, to reproduce their kind, to evolve complex organisms or cells, to exercise germinative affinity, &c.

If then (as was before said at the end of Chapter VIII.) such innate powers must be attributed to chemical atoms, to mineral species, to gemmules, and to physiological units, it is only reasonable to attribute such to each individual organism.

The conception of such internal and latent capabilities is somewhat like that of Mr. Galton, before mentioned, according to which the organic world consists of entities, each of which is, as it were, a spheroid with many facets on its surface, upon one of which it reposes in stable equilibrium. When by the accumulated action of incident forces this equilibrium is disturbed, the spheroid is supposed to turn over until it settles on an adjacent facet once more in stable equilibrium.

The internal tendency of an organism to certain considerable and definite changes would correspond to the facets on the surface of the spheroid.

It may be objected that we have no knowledge as to how terrestrial, cosmical and other forces can affect organisms so as to stimulate and evolve these latent, merely potential forms. But we have had evidence that such mysterious agencies *do* affect organisms in ways as yet inexplicable, in the very remarkable effects of geographical conditions which were detailed in the third chapter.

It is quite conceivable that the material organic world may be so constituted that the simultaneous action upon it of all known forces, mechanical, physical, chemical, magnetic, terrestrial, and cosmical, together with other as yet unknown forces which probably exist, may result in changes which are harmonious and symmetrical, just as the internal nature of vibrating plates causes particles of sand scattered over them to assume definite and symmetrical figures when made to oscillate in different ways by the bow of a violin being drawn along their edges. The results of these combined internal powers and external influences might be represented under the symbol of complex series of vibrations (analogous to those of sound or light) forming a most complex harmony or a display of most varied colours. In such a way the reparation of local injuries might be symbolized as a filling up and completion of an interrupted rhythm. Thus also monstrous aberrations from typical structure might correspond to a discord, and sterility from crossing be compared with the darkness resulting from the interference of waves of light.

Such symbolism will harmonize with the peculiar reproduction, before mentioned, of heads in the body of certain annelids, with the facts of serial homology, as well as those of bilateral and

vertical symmetry. Also, as the atoms of a resonant body may
be made to give out sound by the juxtaposition of a vibrating
tuning-fork, so it is conceivable that the physiological units of a
living organism may be so influenced by surrounding conditions
(organic and other) that the accumulation of these conditions
may upset the previous rhythm of such units, producing modifications in them—a fresh chord in the harmony of nature—a
new species !

But it may be again objected that to say that species arise by
the help of an innate power possessed by organisms is no explanation, but is a reproduction of the absurdity, *l'opium endormit
parcequ'il a une vertu soporifique.* It is contended, however,
that this objection does not apply, even if it be conceded that
there is that force in Molière's ridicule which is generally attributed to it.[1] Much, however, might be said in opposition to
more than one of that brilliant dramatist's smart philosophical
epigrams, just as to the theological ones of Voltaire, or to the
biological one of that other Frenchman who for a time
discredited a cranial skeletal theory by the phrase " Vertèbre
pensante."[2]

In fact, however, it is a real explanation of how a man
lives to say that he lives independently, on his own income,
instead of being supported by his relatives and friends. In the
same way, there is fully as real a distinction between the production of new specific manifestations entirely *ab externo*, and by the
production of the same through an innate force and tendency,

[1] If anyone were to contend that beside the opium there existed a real
distinct objective entity, "its soporific virtue," he would be open to
ridicule indeed. But the constitution of our minds is such that we cannot
but distinguish ideally a thing from its even essential attributes and
qualities. The joke is sufficiently amusing, however, regarded as the
solemn enunciation of a mere truism.

[2] Noticed by Professor Owen in his "Archetype," p. 76. Recently it has
been attempted to discredit Darwinism in France by speaking of it as "*de
la science mousseuse !*"

the determination of which into action is occasioned by external circumstances.

To say that organisms possess this innate power, and that by it new species are from time to time produced, is by no means a mere assertion that they *are* produced, and in an unknown mode. It is the negation of that view which deems external forces alone sufficient, and at the same time the assertion of something positive, to be arrived at by the process of *reductio ad absurdum.*

All physical explanations result ultimately in such conceptions of innate power, or else in that of will force. The far-famed explanation of the celestial motions ends in the conception that every particle of matter has the innate power of attracting every other particle directly as the mass, and inversely as the square of the distance.

We are logically driven to this positive conception if we do not accept the view that there is no force but volition, and that all phenomena whatever are the immediate results of the action of intelligent and self-conscious will.

We have seen that the notion of sudden changes—saltatory actions in nature—has received countenance from Professor Huxley.[1] We must conceive that these jumps are orderly, and according to law, inasmuch as the whole cosmos is such. Such orderly evolution harmonizes with a teleology derived, not indeed from external nature directly, but from the mind of man. On this point, however, more will be said in the next chapter. But, once more, if new species are not manifested by the action of external conditions upon minute indefinite individual differences, in what precise way may we conceive that manifestation to have taken place ?

Are new species now evolving, as they have been from time to time evolved ?   If so, in what way and by what conceivable means ?

[1] " Lay Sermons," p. 342.

In the first place, they must be produced by natural action in pre-existing material, or by supernatural action.

For reasons to be given in the next chapter, the second hypothesis need not be considered.

If, then, new species are and have been evolved from pre-existing material, must that material have been organic or inorganic?

As before said, additional arguments have lately been brought forward to show that individual organisms *do* arise from a basis of *in*-organic material only. As, however, this at the most appears to be the case, if at all, only with the lowest and most minute organisms exclusively, the process cannot be observed, though it may perhaps be fairly inferred.

We may therefore, if for no other reason, dismiss the notion that highly organized animals and plants can be suddenly or gradually built up by any combination of physical forces and natural powers acting externally and internally upon and in merely inorganic material as a base.

But the question is, how have the highest kinds of animals and plants arisen? It seems impossible that they can have appeared otherwise than by the agency of antecedent organisms not greatly different from them.

A multitude of facts, ever increasing in number and importance, all point to such a mode of specific manifestation.

One very good example has been adduced by Professor Flower in the introductory lecture of his first Hunterian Course.[1] It is the reduction in size, to a greater or less degree, of the second and third digits of the foot in Australian marsupials, and this, in spite of the very different form and function of the foot in different groups of those animals.

A similarly significant evidence of relationship is afforded by processes of the zygomatic region of the skull in certain edentates existing and extinct.

[1] Introductory Lecture of February 14, 1870, pp. 24—30, Figs. 1—4. Churchill and Sons.)

Again, the relation between existing and recent faunas of the different regions of the world, and the predominating (though by no means exclusive) march of organization, from the more general to the more special, point in the same direction.

Almost all the facts brought forward by the patient industry of Mr. Darwin in support of his theory of " Natural Selection," are of course available as evidence in favour of the agency of pre-existing and similar animals in specific evolution.

Now the new forms must be produced by changes taking place in organisms in, after or before their birth, either in their embryonic, or towards or in their adult, condition.

Examples of strange births are sufficiently common, and they may arise either from direct embryonic modifications or apparently from some obscure change in the parental action. To the former category belong the hosts of instances of malformation through arrest of development, and perhaps generally monstrosities of some sort are the result of such affections of the embryo. To the second category belong all cases of hybridism, of cross breed, and in all probability the new varieties and forms, such as the memorable one of the black-shouldered peacock. In all these cases we do not have abortions or monstrosities, but more or less harmonious forms often of great functional activity, endowed with marked viability and generative prepotency, except in the case of hybrids, when we often find even a more marked generative impotency.

It seems probable therefore that new species may arise from some constitutional affection of parental forms—an affection mainly, if not exclusively, of their generative system. Mr. Darwin has carefully collected[1] numerous instances to show how excessively sensitive to various influences this system is. He says : [2] " Sterility is independent of general health, and is often

---

[1] See especially "Animals and Plants under Domestication," vol. ii. chap. xviii.   [2] " Origin of Species," 5th edition, pp. 323, 324.

accompanied by excess of size, or great luxuriance," and, "No one can tell, till he tries, whether any particular animal will breed under confinement, or any exotic plant seed freely under culture." Again, "When a new character arises, whatever its nature may be, it generally tends to be inherited, at least in a temporary and sometimes in a most persistent manner." [1] Yet the obscure action of conditions will alter characters long inherited, as the grandchildren of Aylesbury ducks, removed to a distant part of England, completely lost their early habit of incubation, and hatched their eggs at the same time with the common ducks of the same place." [2]

Mr. Darwin quotes Mr. Bartlett as saying: "It is remarkable that lions breed more freely in travelling collections than in the zoological gardens ; probably the constant excitement and irritation produced by moving from place to place, or change of air, may have considerable influence in the matter." [3]

Mr. Darwin also says: "There is reason to believe that insects are affected by confinement like the higher animals," and he gives examples. [4]

Again, he gives examples of change of plumage in the linnet, bunting, oriole, and other birds, and of the temporary modification of the horns of a male deer during a voyage. [5]

Finally, he adds that these changes cannot be attributed to loss of health or vigour, " when we reflect how healthy, long-lived, and vigorous many animals are under captivity, such as parrots, and hawks when used for hawking, chetahs when used for hunting, and elephants. The reproductive organs themselves are not diseased ; and the diseases from which animals in menageries usually perish, are not those which in any way affect their fertility. No domestic animal is more subject to disease than the sheep, yet it is remarkably prolific. . . . . It would appear

---

[1] " Animals and Plants under Domestication," vol. ii. p. 2.
[2] Ibid. p. 25.   [3] Ibid. p. 151.
[4] Ibid. p. 157.   [5] Ibid. p. 158.

that any change in the habits of life, whatever these habits may be, if great enough, tends to affect in an inexplicable manner the powers of reproduction."

Such, then, is the singular sensitiveness of the generative system.

As to the means by which that system is affected, we see that a variety of conditions affect it; but as to the modes in which they act upon it, we have as yet little if any clue.

We have also seen the singular effects (in tailed Lepidoptera, &c.) of causes connected with geographical distribution, the mode of action of which is as yet quite inexplicable ; and we have also seen the innate tendency which there appears to be in certain groups (birds of paradise, &c.) to develop peculiarities of a special kind.

It is, to say the least, probable that other influences exist, terrestrial and cosmical, as yet un-noted. The gradually accumulating or diversely combining actions of all these on highly sensitive structures, which are themselves possessed of internal responsive powers and tendencies, may well result in occasional repeated productions of forms harmonious and vigorous, and differing from the parental forms in proportion to the result of the combining or conflicting action of all external and internal influences.

If, in the past history of this planet, more causes ever intervened, or intervened more energetically than at present, we might *a priori* expect a richer and more various evolution of forms more radically differing than any which could be produced under conditions of more perfect equilibrium. At the same time, if it be true that the last few thousand years have been a period of remarkable and exceptional uniformity as regards this planet's astronomical relations, there are then some grounds for thinking that organic evolution may have been exceptionally depressed during the same epoch.

Now, as to the fact that sudden changes and sudden developments have occurred, and as to the probability that such changes are likely to occur, evidence was given in Chapter IV.

In Chapter V. we also saw that minerals become modified suddenly and considerably by the action of incident forces—as, *e.g.*, the production of hexagonal tabular crystals of carbonate of copper by sulphuric acid, and of long rectangular prisms by ammonia, &c.

We have thus a certain antecedent probability that if changes are produced in specific manifestation through incident forces, these changes will be sensible and considerable, not minute and infinitesimal.

Consequently, it is probable that new species have appeared from time to time with comparative suddenness, and that they still continue so to arise if all the conditions necessary for specific evolution now obtain.

This probability will be increased if the observations of Dr. Bastian are confirmed by future investigation. According to his report, when the requisite conditions were supplied, the transformations which appeared to take place (from very low to higher organisms) were sudden, definite, and complete.

Therefore, if this is so, there must probably exist in higher forms a similar tendency to such change. That tendency may indeed be long suppressed, and ultimately modified by the action of heredity—an action which would increase in force with the increase in the perfection and complexity of the organism affected. Still we might expect that such changes as do take place would be also sudden, definite, and complete.

Moreover, as the same causes produce the same effects, several individual parent forms must often have been similarly and simultaneously affected. That they should be so affected—at least that several similarly modified individuals should simultaneously arise—has been seen to be a generally necessary circumstance for the permanent duration of such new modifications.

It is also conceivable that such new forms may be endowed

with excessive constitutional strength and viability, and with generative prepotency, as was the case with the black-shouldered
peacock in Sir J. Trevelyan's flock.   This flock was entirely composed of the common kind, and yet the new form rapidly developed itself " *to the extinction of the previously existing breed.*"[1]

Indeed, the notion accepted by both Mr. Darwin and Mr.
Herbert Spencer, and which is plainly the fact (namely, that
changes of conditions and incident forces, within limits, augment
the viability and fertility of individuals), harmonizes well with
the suggested possibility as to an augmented viability and prepotency in new organic forms evolved by peculiar consentaneous
actions of conditions and forces, both external and internal.

The remarkable series of changes noted by Dr. Bastian were
certainly not produced by external incident forces *only*, but by
these acting on a peculiar *materia*, having special properties and
powers.   Therefore, the changes were induced by the consentaneous action of internal and external forces.[2]   In the same way
then, we may expect changes in higher forms to be evolved by
similar united action of internal and external forces.

One other point may here be alluded to.   When the remarkable way in which structure and function simultaneously change,
is borne in mind ; when those numerous instances in which
nature has supplied similar wants by similar means, as detailed
in Chapter III., are remembered; when also all the wonderful
contrivances of orchids, of mimicry, and the strange complexity
of certain instinctive actions are considered : then the conviction
forces itself on many minds that the organic world is the
expression of an intelligence of some kind.   This view has
been well advocated by Mr. Joseph John Murphy, in his recent
work so often here referred to.

[1] " Animals and Plants under Domestication," vol. i. p. 291.
[2] Though hardly necessary, it may be well to remark that the views here
advocated in no way depend upon the truth of the doctrine of Spontaneous
Generation.

This intelligence, however, is evidently not altogether such as ours, or else has other ends in view than those most obvious to us. For the end is often attained in singularly roundabout ways, or with a prodigality of means which seems out of all proportion with the result : not with the simple action directed to one end which generally marks human activity.

Organic nature then speaks clearly to many minds of the action of an intelligence resulting, on the whole and in the main, in order, harmony, and beauty, yet of an intelligence the ways of which are not such as ours.

This view of evolution harmonizes well with Theistic conceptions ; not, of course, that this harmony is brought forward as an argument in its favour generally, but it will have weight with those who are convinced that Theism reposes upon solid grounds of reason as *the* rational view of the universe. To such it may be observed that, thus conceived, the Divine action has that slight amount of resemblance to, and that wide amount of divergence from what human action would be, which might be expected *a priori*—might be expected, that is, from a Being whose nature and aims are utterly beyond our power to imagine, however faintly, but whose truth and goodness are the fountain and source of our own perceptions of such qualities.

The view of evolution maintained in this work, though arrived at in complete independence, yet seems to agree in many respects with the views advocated by Professor Owen in the last volume of his " Anatomy of Vertebrates," under the term "derivation." He says :[1] " Derivation holds that every species changes in time, by virtue of inherent tendencies thereto. 'Natural Selection' holds that no such change can take place without the influence of altered external circumstances.[2]   'De-

---

[1] Vol. iii. p. 808.

[2] This is hardly an exact representation of Mr. Darwin's view. On his theory, if a favourable variation happens to arise (the external circumstances remaining the same), it will yet be preserved.

rivation' sees among the effects of the innate tendency to change irrespective of altered circumstances, a manifestation of creative power in the variety and beauty of the results; and, in the ultimate forthcoming of a being susceptible of appreciating such beauty, evidence of the pre-ordaining of such relation of power to the appreciation. 'Natural Selection' acknowledges that if ornament or beauty, in itself, should be a purpose in creation, it would be absolutely fatal to it as a hypothesis."

"'Natural Selection' sees grandeur in the view of life, with its several powers, having been originally breathed by the Creator into a few forms or into one. 'Derivation' sees therein a narrow invocation of a special miracle and an unworthy limitation of creative power, the grandeur of which is manifested daily, hourly, in calling into life many forms, by conversion of physical and chemical into vital modes of force, under as many diversified conditions of the requisite elements to be so combined."

The view propounded in this work allows, however, a greater and more important part to the share of external influences, it being believed by the Author, however, that these external influences equally with the internal ones are the results of one harmonious action underlying the whole of nature, organic and inorganic, cosmical, physical, chemical, terrestrial, vital, and social.

According to this view, an internal law presides over the actions of every part of every individual, and of every organism as a unit, and of the entire organic world as a whole. It is believed that this conception of an internal innate force will ever remain necessary, however much its subordinate processes and actions may become explicable:

That by such a force, from time to time, new species are manifested by ordinary generation just as *Pavo nigripennis* appeared suddenly, these new forms not being monstrosities but harmonious self-consistent wholes. That thus, as specific distinctness is manifested by obscure sexual conditions, so in obscure sexual modifications specific distinctions arise.

That these "jumps" are considerable in comparison with the minute variations of "Natural Selection"—are in fact sensible steps, such as discriminate species from species.

That the latent tendency which exists to these sudden evolutions is determined to action by the stimulus of external conditions.

That "Natural Selection" rigorously destroys monstrosities, and abortive and feeble attempts at the performance of the evolutionary process.

That "Natural Selection" removes the antecedent species rapidly when the new one evolved is more in harmony with surrounding conditions.

That "Natural Selection" favours and develops useful variations, though it is impotent to originate them or to erect the physiological barrier which seems to exist between species.

By some such conception as this, the difficulties here enumerated, which beset the theory of "Natural Selection" pure and simple, are to be got over.

Thus, for example, the difficulties discussed in the first chapter—namely, those as to the origins and first beginnings of certain structures—are completely evaded.

Again, as to the independent origin of closely similar structures, such as the eyes of the Vertebrata and cuttle-fishes, the difficulty is removed if we may adopt the conception of an innate force similarly directed in each case, and assisted by favourable external conditions.

Specific stability, limitation to variability, and the facts of reversion, all harmonize with the view here put forward. The same may be said with regard to the significant facts of homology, and of organic symmetry ; and our consideration of the hypothesis of Pangenesis in Chapter X., has seemed to result in a view as to innate powers which accords well with what is here advocated.

The evolutionary hypothesis here advocated also serves to explain all those remarkable facts which were stated in the first chapter to be explicable by the theory of Natural Selection, namely, the relation of existing to recent faunas and floras; the phenomena of homology and of rudimentary structures; also the processes gone through in development; and lastly, the wonderful facts of mimicry.

Finally, the view adopted is the synthesis of many distinct and, at first sight, conflicting conceptions, each of which contains elements of truth, and all of which it appears to be able more or less to harmonize.

Thus it has been seen that "Natural Selection" is accepted. It acts and must act, though alone it does not appear capable of fulfilling the task assigned to it by Mr. Darwin.

Pangenesis has probably also much truth in it, and has certainly afforded valuable and pregnant suggestions, but un-aided and alone it seems inadequate to explain the evolution of the individual organism.

Those three conceptions of the organic world which may be spoken of as the teleological, the typical, and the transmutationist, have often been regarded as mutually antagonistic and conflicting.

The genesis of species as here conceived, however, accepts, locates, and harmonizes all the three.

Teleology concerns the ends for which organisms were designed. The recognition, therefore, that their formation took place by an evolution not fortuitous, in no way invalidates the acknowledgment of their final causes if on other grounds there are reasons for believing that such final causes exist.

Conformity to type, or the creation of species according to certain "divine ideas," is in no way interfered with by such a process of evolution as is here advocated. Such "divine ideas" must be accepted or declined upon quite other grounds than the mode of their realization, and of their manifestation in the world of sensible phenomena.

Transmutationism (an old name for the evolutionary hypothesis), which was conceived at one time to be the very antithesis to the two preceding conceptions, harmonizes well with them if the evolution be conceived to be orderly and designed. It will in the next chapter be shown to be completely in harmony with conceptions, upon the acceptance of which "final causes" and "divine ideal archetypes" alike depend.

Thus then, if the cumulative argument put forward in this book is valid, we must admit the insufficiency of Natural Selection both on account of the residuary phenomena it fails to explain, and on account of certain other phenomena which seem actually to conflict with that theory. We have seen that though the laws of nature are constant, yet some of the conditions which determine specific change may be exceptionally absent at the present epoch of the world's history; also that it is not only possible, but highly probable, that an internal power or tendency is an important if not the main agent in evoking the manifestation of new species on the scene of realized existence, and that in any case, from the facts of homology, innate internal powers to the full as mysterious must anyhow be accepted, whether they act in specific origination or not. Besides all this, we have seen that it is probable that the action of this innate power is stimulated, evoked, and determined by external conditions, and also that the same external conditions, in the shape of "Natural Selection," play an important part in the evolutionary process : and finally, it has been affirmed that the view here advocated, while it is supported by the facts on which Darwinism rests, is not open to the objections and difficulties which oppose themselves to the reception of "Natural Selection," as the exclusive or even as the main agent in the successive and orderly evolution of organic forms in the *genesis of species.*

CHAPTER XII.

THEOLOGY AND EVOLUTION.

Prejudiced opinions on the subject.—"Creation" sometimes denied from prejudice.—The unknowable.—Mr. Herbert Spencer's objections to theism ; to creation.—Meanings of term "creation."—Confusion from not distinguishing between "primary" and "derivative" creation.—Mr. Darwin's objections.—Bearing of Christianity on the theory of evolution.—Supposed opposition, the result of a misconception.—Theological authority not opposed to evolution.—St. Augustin.—St. Thomas Aquinas.—Certain consequences of want of flexibility of mind.—Reason and imagination.—The first cause and demonstration.—Parallel between Christianity and natural theology.—What evolution of species is.—Professor Agassiz.—Innate powers must be recognized.—Bearing of evolution on religious belief.—Professor Huxley.—Professor Owen.—Mr. Wallace.—Mr. Darwin.—*A priori* conception of Divine action.—Origin of man.—Absolute creation and dogma.—Mr. Wallace's view.—A supernatural origin for man's body not necessary.—Two orders of being in man.—Two modes of origin.—Harmony of the physical, hyperphysical, and supernatural.—Reconciliation of science and religion as regards evolution.—Conclusion.

THE special "Darwinian Theory" and that of an evolutionary process neither excessively minute nor fortuitous, having now been considered, it is time to turn to the important question, whether both or either of these conceptions may have any bearing, and if any, what, upon Christian belief?

Some readers will consider such an inquiry to be a work of supererogation. Seeing clearly themselves the absurdity of prevalent popular views, and the shallowness of popular objections, they may be impatient of any discussion on the subject.   But it is submitted that there are many minds worthy of the highest esteem and of every consideration, which have regarded the subject hitherto almost exclusively from one point of view ; that there are some persons who are opposed to the progress (in their own minds or in that of their children or dependents) of physical scientific truth—the natural revelation— through a mistaken estimate of its religious bearings, while there are others who are zealous in its promotion from a precisely similar error.   For the sake of both these then the Author may perhaps be pardoned for entering slightly on very elementary matters relating to the question, whether evolution or Darwinism have any, and if any, what, bearing on theology ?

There are at least two classes of men who will certainly assert that they have a very important and highly significant bearing upon it.

One of these classes consists of persons zealous for religion indeed, but who identify orthodoxy with their own private interpretation of Scripture or with narrow opinions in which they have been brought up—opinions doubtless widely spread, but at the same time destitute of any distinct and authoritative sanction on the part of the Christian Church.

The other class is made up of men hostile to religion, and who are glad to make use of any and every argument which they think may possibly be available against it.

Some individuals within this latter class may not believe in the existence of God, but may yet abstain from publicly avowing this absence of belief, contenting themselves with denials of " creation" and " design," though these denials are really consequences of their attitude of mind respecting the most important and fundamental of all beliefs.

Without a distinct belief in a personal God it is impossible to have any religion worthy of the name, and no one can at the same time accept the Christian religion and deny the dogma of creation.

" I believe in God," "the Creator of Heaven and Earth," the very first clauses of the Apostles' Creed, formally commit those who accept them to the assertion of this belief. If, therefore, any theory of physical science really conflicts with such an authoritative statement, its importance to Christians is unquestionable.

As, however, " creation" forms a part of " revelation," and as " revelation" appeals for its acceptance to " reason" which has to prepare a basis for it by an intelligent acceptance of theism on *purely rational grounds*, it is necessary to start with a few words as to the reasonableness of belief in God, which indeed are less superfluous than some readers may perhaps imagine ; " a few words," because this is not the place where the argument can be drawn out, but only one or two hints given in reply to certain modern objections.

No better example perhaps can be taken, as a type of these objections, than a passage in Mr. Herbert Spencer's First Principles.[1] This author constantly speaks of the " ultimate cause of things " as " the Unknowable," a term singularly unfortunate, and as Mr. James Martineau has pointed out,[2] even self-

---

[1] See 2nd edition, p. 113.

[2] " Essays, Philosophical and Theological," Trübner and Co., First Series, 1866, p. 190. "Every relative disability may be read two ways. A disqualification in the nature of thought for knowing x is, from the other side, a disqualification in the nature of x from being known. To say then that the First Cause is wholly removed from our apprehension is not simply a disclaimer of faculty on our part : it is a charge of inability against the First Cause too. The dictum about it is this : ' It is a Being that may exist out of knowledge, but that is precluded from entering within the sphere of knowledge.' We are told in one breath that this Being must be in every sense ' perfect, complete, total—including in itself all power, and transcending all law' (p. 38); and in another that this perfect and omnipotent One is totally incapable of revealing any one of an infinite store of attributes. Need we point out the contradictions which this position involves ? If you abide by it, you deny the Absolute and Infinite in the very act of affirming it, for, in debarring the First Cause from self-revelation, you

contradictory : for that entity, the knowledge of the existence of which presses itself ever more and more upon the cultivated intellect, cannot be the unknown, still less *the unknowable*, because we certainly know it, in that we know for certain that it exists. Nay more, to predicate incognoscibility of it, is even a certain knowledge of the mode of its existence. Mr. H. Spencer says:[1] "The consciousness of an Inscrutable Power manifested to us through all phenomena has been growing ever clearer; and must eventually be freed from its imperfections. The certainty that on the one hand such a Power exists, while on the other hand its nature transcends intuition, and is beyond imagination, is the certainty towards which intelligence has from the first been progressing." One would think then that the familiar and accepted word "the Inscrutable" (which is in this passage actually employed, and to which no theologian would object) would be an indefinitely better term than "the unknowable." The above extract has, however, such a theistic aspect that some readers may think the opposition here offered superfluous; it may be well, therefore, to quote two other sentences. In another place he observes,[2] "Passing over the consideration of credibility, and confining ourselves to that of conceivability, we see that atheism, pantheism, and theism, when rigorously analysed, severally prove to be absolutely unthinkable ;" and speaking of "every form of religion," he adds,[3] "The analysis of every possible hypothesis proves, not simply that no hypothesis is sufficient but that no hypothesis is even thinkable." The unknowable is admitted to be a power which cannot be regarded as having sympathy with us, but as one to which no emotion whatever

impose a limit on its nature. And in the very act of declaring the First Cause incognizable, you do not permit it to remain unknown. For that only is unknown, of which you can neither affirm nor deny any predicate ; here you deny the power of self-disclosure to the 'Absolute,' of which therefore something is known ;—viz., that nothing can be known !"

[1] Loc. cit. p. 108.   [2] Loc. cit. p. 43.   [3] Loc. cit. p. 46.

can be ascribed, and we are expressly forbidden "by *duty*," to affirm personality of God as much as to deny it of Him. How such a being can be presented as an object on which to exercise religious emotion it is difficult indeed to understand.[1] Aspiration, love, devotion to be poured forth upon what we can never know, upon what we can never affirm to know, or care for, us, our thoughts or actions, or to possess the attributes of wisdom and goodness! The worship offered in such a religion must be, as Professor Huxley says,[2] "for the most part of the silent sort"—silent not only as to the spoken word, but silent as to the mental conception also. It will be difficult to distinguish the follower of this religion from the follower of none, and the man who declines either to assert or to deny the existence of God, is practically in the position of an atheist. For theism enjoins the cultivation of sentiments of love and devotion to God, and the practice of their external expression. Atheism forbids both, while the simply non-theist abstains in conformity with the prohibition of the atheist and thus practically sides with him. Moreover, since man cannot imagine that of which he has no experience in any way whatever, and since he has experience only of *human* perfections and of the powers and properties of *inferior* existences ; if he be required to deny human perfections and to abstain from making use of such conceptions, he is thereby necessarily reduced to others of an inferior order.

[1] Mr. J. Martineau, in his "Essays," vol. i. p. 211, observes, "Mr. Spencer's conditions of pious worship are hard to satisfy ; there must be between the Divine and human no communion of thought, relations of conscience, or approach of affection." . . . , "But you cannot constitute a religion out of mystery alone, any more than out of knowledge alone ; nor can you measure the relation of doctrines to humility and piety by the mere amount of conscious darkness which they leave. All worship, being directed to what is *above* us and transcends our comprehension, stands in presence of a mystery. But not all that stands before a mystery is worship."

[2] "Lay Sermons," p. 20.

Mr. H. Spencer says,[1] " Those who espouse this alternative position, make the erroneous assumption that the choice is between personality and something lower than personality; whereas the choice is rather between personality and something higher. Is it not just possible that there is a mode of being as much transcending intelligence and will, as these transcend mechanical motion ? "

" It is true we are totally unable to conceive any such higher mode of being. But this is not a reason for questioning its existence; it is rather the reverse." " May we not therefore rightly refrain from assigning to the ' ultimate cause ' any attributes whatever, on the ground that such attributes, derived as they must be from our own natures, are not elevations but degradations ? " The way however to arrive at the object aimed at (*i.e.* to obtain the best attainable conception of the First Cause) is not to refrain from *the only conceptions possible to us*, but to seek the very highest of these, and then declare their utter inadequacy; and this is precisely the course which has been pursued by theologians. It is to be regretted that before writing on this matter Mr. Spencer did not more thoroughly acquaint himself with the ordinary doctrine on the subject. It is always taught in the Church schools of divinity, that nothing, not even *existence*, is to be predicated *univocally* of " God " and " creatures;" that after exhausting ingenuity to arrive at the loftiest possible conceptions, we must declare them to be *utterly inadequate;* that, after all, they are but accommodations to human infirmity; that they are in a sense objectively false (because of their inadequacy), though subjectively and very practically true. But the difference between this mode of treatment and that adopted by Mr. Spencer is wide indeed; for the practical result of the mode inculcated by the Church is that each one may freely affirm and act upon

[1] Loc. cit. p. 109.

the highest human conceptions he can attain of the power, wisdom, and goodness of God, His watchful care, His loving providence for every man, at every moment and in every need ; for the Christian knows that the falseness of his conceptions lies only in their *inadequacy;* he may therefore strengthen and refresh himself, may rejoice and revel in conceptions of the goodness of God, drawn from the tenderest human images of fatherly care and love, or he may chasten and abase himself by consideration of the awful holiness and unapproachable majesty of the Divinity derived from analogous sources, knowing that no thought of man can ever be *true enough,* can ever attain the incomprehensible reality, which nevertheless really *is* all that can be conceived, *plus* an inconceivable infinity beyond.

A good illustration of what is here meant, and of the difference between the theistic position and Mr. Spencer's, may be supplied by an example he has himself proposed. Thus,[1] he imagines an intelligent watch speculating as to its maker, and conceiving of him in terms of watch-being, and figuring him as furnished with springs, escapements, cogged wheels, &c., his motions facilitated by oil—in a word, like himself. It is assumed by Mr. Spencer that this necessary watch conception would be completely false, and the illustration is made use of to show "the presumption of theologians"—the absurdity and unreasonableness of those men who figure the incomprehensible cause of all phenomena as a Being in some way comparable with man. Now, putting aside for the moment all other considerations, and accepting the illustration, surely the example demonstrates rather the unreasonableness of the *objector himself!* It is true, indeed, that a man is an organism indefinitely more complex and perfect than any watch ; but if the watch could only conceive of its maker in watch terms, or else in terms altogether inferior, the watch would plainly be right in speaking of

[1] Loc. cit. p. 111.

its maker as a, to it, inconceivably perfect kind of watch, acknowledging at the same time, that this, its conception of him, was *utterly inadequate*, although the best its inferior nature allowed it to form.　For if, instead of so conceiving of its maker, it refused to make use of these relative perfections as a makeshift, and so necessarily thought of him as amorphous metal, or mere oil, or by the help of any other inferior conception which a watch might be imagined capable of entertaining, that watch would be wrong indeed.　For man can much more properly be compared with, and has much more affinity to, a perfect watch in full activity than to a mere piece of metal, or drop of oil.　But the watch is even more in the right still, for its maker, man, virtually *has* the cogged wheels, springs, escapements, oil, &c., which the watch's conception has been supposed to attribute to him; inasmuch as all these parts must have existed as distinct ideas in the human watchmaker's mind before he could actually construct the clock formed by him.　Nor is even this all, for, by the hypothesis, the watch *thinks*.　It must, therefore, think of its maker as "a thinking being," and in this it is *absolutely and completely right.*[1]　Either, therefore, the hypothesis is *absurd* or it actually *demonstrates the very position it was chosen to refute.* Unquestionably, then, on the mere ground taken by Mr. Herbert Spencer himself, if we are compelled to think of the First Cause either in human terms (but with human imperfections abstracted and human perfections carried to the highest conceivable degree), or, on the other hand, in terms decidedly inferior, such as those are driven to who think of Him, but decline to accept as a help the term "personality;" there can be no question but that the first conception is immeasurably nearer the truth than the second.　Yet the latter is the one put forward and advocated by that author in spite of its unreasonableness, and in spite also

---

[1] In this criticism on Mr. Herbert Spencer, the Author finds he has been anticipated by Mr. James Martineau. (See "Essays," vol. i. p. 208.)

of its conflicting with the whole moral nature of man and all his noblest aspirations.

Again, Mr. Herbert Spencer objects to the conception of God as "first cause," on the ground that "when our symbolic conceptions are such that no cumulative or indirect processes of thought can enable us to ascertain that there are corresponding actualities, nor any predictions be made whose fulfilment can prove this, then they are altogether vicious and illusive, and in no way distinguishable from pure fictions." [1]

Now, it is quite true that "symbolic conceptions," which are not to be justified either (1) by presentations of sense, or (2) by intuitions, are invalid as representations of real truth. Yet the conception of God referred to *is* justified by our primary intuitions, and we can assure ourselves that it *does* stand for an actuality by comparing it with (1) our intuitions of free-will and causation, and (2) our intuitions of morality and responsibility. That we *have* these intuitions is a point on which the Author joins issue with Mr. Spencer, and confidently affirms that they cannot logically be denied without at the same time complete and absolute scepticism resulting from such denial—scepticism wherein vanishes any certainty as to the existence both of Mr. Spencer and his critic, and by which it is equally impossible to have a thought free from doubt, or to go so far as to affirm the existence of that very doubt or of the doubter who doubts it.

It may not be amiss here to protest against the intolerable assumption of a certain school, who are continually talking in lofty terms of "science," but who actually speak of primary religious conceptions as "unscientific," and habitually employ the word "science," when they should limit it by the prefix "physical." This is the more amazing as not a few of this school adopt the idealist philosophy, and affirm that "matter and force" are but names for certain "modes of consciousness." It might be expected of them at least to admit that opinions which repose

[1] Loc. cit. p. 29.

on primary and fundamental intuitions, are especially and *par excellence* scientific.

Such are some of the objections to the Christian conception of God. We may now turn to those which are directed against God as the Creator, *i.e.* as the absolute originator of the universe, without the employment of any pre-existing means or material. This is again considered by Mr. Spencer as a thoroughly illegitimate symbolic conception, as much so as the atheistic one —the difficulty as to a *self-existent Creator* being in his opinion equal to that of a *self-existent universe.* To this it may be replied that both are of course equally *unimaginable*, but that it is not a question of facility of conception—not which is easiest to conceive, but which best accounts for, and accords with, psychological facts; namely, with the above-mentioned intuitions. It is contended that *we have* these primary intuitions, and that with these the conception of a self-existent Creator is perfectly harmonious. On the other hand, the notion of a self-existent universe—that there is no real distinction between the finite and the infinite—that the universe and ourselves are one and the same things with the infinite and the self-existent; these assertions, in *addition to* being unimaginable, *contradict* our primary intuitions.

Mr. Darwin's objections to " Creation " are of quite a different kind, and, before entering upon them, it will be well to endeavour clearly to understand what we mean by " Creation," in the various senses in which the term may be used.

In the strictest and highest sense " Creation " is the absolute origination of anything by God without pre-existing means or material, and is a *supernatural* act.[1]

In the secondary and lower sense, " Creation " is the formation of anything by God *derivatively;* that is, that the preceding matter has been created with the potentiality to evolve from it,

---

[1] The Author means by this, that it is *directly* and *immediately* the act of God, the word " supernatural " being used in a sense convenient for the purposes of this work, and not in its ordinary theological sense.

under suitable conditions, all the various forms it subsequently assumes. And this power having been conferred by God in the first instance, and those laws and powers having been instituted by Him, through the action of which the suitable conditions are supplied, He is said in this lower sense to create such various subsequent forms. This is the *natural* action of God in the physical world, as distinguished from His direct, or, as it may be here called, supernatural action.

In yet a third sense, the word " Creation " may be more or less improperly applied to the construction of any complex formation or state by a voluntary self-conscious being who makes use of the powers and laws which God has imposed, as when a man is spoken of as the creator of a museum, or of " his own fortune," &c. Such action of a created conscious intelligence is purely natural, but more than physical, and may be conveniently spoken of as hyperphysical.

We have thus (1) direct or supernatural action ; (2) physical action; and (3) hyperphysical action—the two latter both belonging to the order of nature.[1] Neither the physical nor the hyperphysical actions, however, exclude the idea of the Divine concurrence, and with every consistent theist that idea is necessarily included. Dr. Asa Gray has given expression to this.[2] He says, " Agreeing that plants and animals were produced by Omnipotent fiat, does not exclude the idea of natural order and what we call secondary causes. The record of the fiat —' Let the earth bring forth grass, the herb yielding seed,' &c., ' let the earth bring forth the living creature after his kind '—seems even to imply them," and leads to the conclusion that the various kinds were produced through natural agencies.

[1] The phrase " order of nature " is not here used in its theological sense as distinguished from the " order of grace," but as a term, here convenient, to denote actions not due to direct and immediate Divine intervention.

[2] " A Free Examination of Darwin's Treatise," p. 29, reprinted from the *Atlantic Monthly* for July, August, and October, 1860.

Now, much confusion has arisen from not keeping clearly in view this distinction between *absolute* creation and *derivative* creation. With the first, physical science has plainly nothing whatever to do, and is impotent to prove or to refute it. The second is also safe from any attack on the part of physical science, for it is primarily derived from psychical not physical phenomena. The greater part of the apparent force possessed by objectors to creation, like Mr. Darwin, lies in their treating the assertion of derivative creation as if it was an assertion of absolute creation, or at least of supernatural action. Thus, he asks whether some of his opponents believe " that at innumerable periods in the earth's history, certain elemental atoms have been commanded suddenly to flash into living tissues."[1] Certain of Mr. Darwin's objections, however, are not physical, but *metaphysical,* and really attack the dogma of secondary or derivative creation, though to some perhaps they may appear to be directed against absolute creation only.

Thus he uses, as an illustration, the conception of a man who builds an edifice from fragments of rock at the base of a precipice, by selecting for the construction of the various parts of the building the pieces which are the most suitable owing to the shape they happen to have broken into. Afterwards, alluding to this illustration, he says,[2] "The shape of the fragments of stone at the base of our precipice may be called accidental, but this is not strictly correct, for the shape of each depends on a long sequence of events, all obeying natural laws, on the nature of the rock, on the lines of stratification or cleavage, on the form of the mountain which depends on its upheaval and subsequent denudation, and lastly, on the storm and earthquake which threw down the fragments. But in regard to the use to which the fragments may be put, their shape may strictly be

[1] "Origin of Species," 5th edition, p. 571.
[2] "Animals and Plants under Domestication," vol. ii. p. 431.

said to be accidental. And here we are led to face a great difficulty, in alluding to which I am aware that I am travelling beyond my proper province."

" An omniscient Creator must have foreseen every consequence which results from the laws imposed by Him ; but can it be reasonably maintained that the Creator intentionally ordered, if we use the words in any ordinary sense, that certain fragments of rock should assume certain shapes, so that the builder might erect his edifice ? If the various laws which have determined the shape of each fragment were not predetermined for the builder's sake, can it with any greater probability be maintained that He specially ordained, for the sake of the breeder, each of the innumerable variations in our domestic animals and plants— many of these variations being of no service to man, and not beneficial, far more often injurious, to the creatures themselves ? Did He ordain that the crop and tail-feathers of the pigeon should vary, in order that the fancier might make his grotesque pouter and fantail breeds ? Did He cause the frame and mental qualities of the dog to vary, in order that a breed might be formed of indomitable ferocity, with jaws fitted to pin down the bull for man's brutal sport ? But, if we give up the principle in one case —if we do not admit that the variations of the primeval dog were intentionally guided, in order that the greyhound, for instance, that perfect image of symmetry and vigour, might be formed,—no shadow of reason can be assigned for the belief that the variations, alike in nature, and the result of the same general laws, which have been the groundwork through Natural Selec- tion of the formation of the most perfectly adapted animals in the world, man included, were intentionally and specially guided. However much we may wish it, we can hardly follow Professor Asa Gray in his belief that 'variation has been led along certain beneficial lines,' like a stream 'along definite and useful lines of irrigation.' "

" If we assume that each particular variation was from the

beginning of all time pre-ordained, the plasticity of the organization, which leads to many injurious deviations of structure, as
well as that redundant power of reproduction which inevitably
leads to a struggle for existence, and, as a consequence, to the
Natural Selection and survival of the fittest, must appear to us
superfluous laws of nature. On the other hand, an omnipotent
and omniscient Creator ordains everything and foresees everything. Thus we are brought face to face with a difficulty as
insoluble as is that of freewill and predestination."

Before proceeding to reply to this remarkable passage, it
may be well to remind some readers that belief in the existence of God, in His primary creation of the universe, and in
His derivative creation of all kinds of being, inorganic and
organic, do not repose upon physical phenomena, but, as has
been said, on primary intuitions. To deny or ridicule any of
these beliefs on physical grounds is to commit the fallacy of
ignoratio elenchi. It is to commit an absurdity analogous to that
of saying a blind child could not recognize his father because he
could not *see* him, forgetting that he could *hear* and *feel* him.
Yet there are some who appear to find it unreasonable and absurd
that men should regard phenomena in a light not furnished by
or deducible from the very phenomena themselves, although the
men so regarding them avow that the light in which they do
view them comes from quite another source. It is as if a man,
A, coming into B's room and finding there a butterfly, should
insist that B had no right to believe that the butterfly had not
flown in at the open window, inasmuch as there was nothing
about the room or insect to lead to any other belief; while B
can well sustain his right so to believe, he having met C, who
told him he brought in the chrysalis and, having seen the insect
emerge, took away the skin.

By a similarly narrow and incomplete view the assertion that
human conceptions, such as "the vertebrate idea," &c., are ideas
in the mind of God, is sometimes ridiculed; as if the assertors

either on the one hand pretended to some prodigious acuteness of mind—a far-reaching genius not possessed by most naturalists—or, on the other hand, as if they detected in the very phenomena furnishing such special conception evidences of Divine imaginings. But let the idea of God, according to the highest conceptions of Christianity, be once accepted, and then it becomes simply a truism to say that the mind of the Deity contains all that is *good* and *positive* in the mind of man, *plus*, of course, an absolutely inconceivable infinity beyond. That thus such human conceptions may, nay must, be asserted to be at the same time ideas in the Divine mind also, as every real and separate individual that has been, is, or shall be, is present to the same mind. Nay, more, that such human conceptions are but faint and obscure adumbrations of corresponding ideas which exist in the mind of God in perfection and fulness.[1]

The theist, having arrived at his theistic convictions from quite other sources than a consideration of zoological or botanical

---

[1] The Rev. Baden Powell says, "All sciences approach perfection as they approach to a unity of first principles,—in all cases recurring to or tending towards certain high elementary conceptions which are the representatives of the unity of the great archetypal ideas according to which the whole system is arranged. Inductive conceptions, very partially and imperfectly realized and apprehended by human intellect, are the exponents in our minds of these great principles in nature."

"All science is but the partial reflexion in the *reason of man*, of the great all-pervading *reason of the universe*. And thus the *unity* of science is the reflexion of the *unity* of nature, and of the *unity* of that supreme reason and intelligence which pervades and rules over nature, and from whence all reason and all science is derived." (Unity of Worlds, Essay i., § ii.; Unity of Sciences, pp. 79 and 81.) Also he quotes from Oersted's "Soul in Nature" (pp. 12, 16, 18, 87, 92, and 377). "If the laws of reason did not exist in nature, we should vainly attempt to force them upon her: if the laws of nature did not exist in our reason, we should not be able to comprehend them." . . . . "We find an agreement between our reason and works which our reason did not produce." . . . . "All existence is a dominion of reason." "The laws of nature are laws of reason, and altogether form an endless unity of reason; . . . . one and the same throughout the universe."

phenomena, returns to the consideration of such phenomena and views them in a theistic light without of course asserting or implying that such light has been derived *from them*, or that there is an obligation of reason so to view them on the part of others who refuse to enter upon or to accept those other sources whence have been derived the theistic convictions of the theist.

But Mr. Darwin is not guilty of arguing against metaphysical ideas on physical grounds only, for he employs very distinctly metaphysical ones; namely, his conceptions of the nature and attributes of the First Cause. But what conceptions does he offer us? Nothing but that low anthropomorphism which, unfortunately, he so often seems to treat as the necessary result of Theism. It is again the dummy, helpless and deformed, set up merely for the purpose of being knocked down.

It must once more be insisted on, that though man is indeed compelled to conceive of God in human terms, and to speak of Him by epithets objectively false, from their hopeless inadequacy, yet nevertheless the Christian thinker declares that inadequacy in the strongest manner, and vehemently rejects from his idea of God all terms distinctly implying infirmity or limitation.

Now, Mr. Darwin speaks as if all who believe in the Almighty were compelled to accept as really applicable to the Deity conceptions which affirm limits and imperfections. Thus he says: "Can it be reasonably maintained that the Creator intentionally ordered" "that certain fragments of rock should assume certain shapes, so that the builder might erect his edifice?"

Why, surely every theist must maintain that in the first foundation of the universe—the primary and absolute creation—God saw and knew every purpose which every atom and particle of matter should ever subserve in all suns and systems, and throughout all coming æons of time. It is almost incredible, but nevertheless it seems necessary to think that the difficulty thus proposed rests on a sort of notion that amidst the boundless profusion of nature there is too much for God to superintend; that the number

of objects is too great for an infinite and *omnipresent* being to
attend singly to each and all in their due proportions and needs!
In the same way Mr. Darwin asks whether God can have ordered
the race variations referred to in the passage last quoted, for
the considerations therein mentioned. To this it may be at
once replied that even man often has *several* distinct intentions
and motives for a *single* action, and the theist has no difficulty
in supposing that, out of an infinite number of motives, the
motive mentioned in each case may have been an exceedingly
subordinate one. The theist, though properly attributing to
God what, for want of a better term, he calls "purpose" and
"design," yet affirms that the limitations of human purposes and
motives are by no means applicable to the Divine "purposes." Out
of many, say a thousand million, reasons for the institution of the
laws of the physical universe, some few are to a certain extent
conceivable by us; and amongst these the benefits, material and
moral, accruing from them to men, and to each individual man
in every circumstance of his life, play a certain, perhaps a very
subordinate, part.[1] As Baden Powell observes, "How can we

[1] In the same way Mr. Lewes, in criticising the Duke of Argyll's "Reign
of Law" (*Fortnightly Review*, July 1867, p. 100), asks whether we should
consider that man wise who spilt a gallon of wine in order to fill a wine-
glass? But, because we should not do so, it by no means follows that we
can argue from such an action to the action of God in the visible universe.
For the man's object, in the case supposed, is simply to fill the wine-glass,
and the wine spilt is so much loss. With God it may be entirely different
in both respects. All these objections are fully met by the principle thus
laid down by St. Thomas Aquinas : " Quod si aliqua causa particularis de-
ficiat a suo effectu, hoc est propter aliquam causam particularem impedi-
antem quæ continetur sub ordine causæ universalis. Unde effectus ordinem
causæ universalis nullo modo potest exire." . . . . " Sicut indigestio contingit
præter ordinem virtutis nutritivæ ex aliquo impedimento, puta ex grossitie
cibi, quam necesse est reducere in aliam causam, et sic usque ad causam
primam universalem. Cum igitur Deus sit prima causa universalis non
unius generi tantum, sed universaliter totius entis, impossibile est quod
aliquid contingat præter ordinem divinæ gubernationis ; sed ex hoc ipso
quod aliquid ex unâ parte videtur exire ab ordine divinæ providentiæ, quo

undertake to affirm, amid all the possibilities of things of which we confessedly know so little, that a thousand ends and purposes may not be answered, because we can trace none, or even imagine none, which seem to our short-sighted faculties to be answered in these particular arrangements?"[1]

The objection to the bull-dog's ferocity in connexion with "man's brutal sport" opens up the familiar but vast question of the existence of evil, a problem the discussion of which would be out of place here. Considering, however, the very great stress which is laid in the present day on the subject of animal suffering by so many amiable and excellent people, one or two remarks on that matter may not be superfluous. To those who accept the belief in God, the soul and moral responsibility; and recognize the full results of that acceptance—to such, physical suffering and moral evil are simply incommensurable. To them the placing of non-moral beings in the same scale with moral agents will be utterly unendurable. But even considering physical pain only, all must admit that this depends greatly on the mental condition of the sufferer. Only during consciousness does it exist, and only in the most highly-organized men does it reach its acme. The Author has been assured that lower races of men appear less keenly sensitive to physical pain than do more cultivated and refined human beings. Thus only in man can there really be any intense degree of suffering, because only in him is there that intellectual recollection of past moments and that anticipation of future ones, which constitute in great part the bitterness of suffering.[2] The momentary pang, the present pain, which beasts endure, though real enough, is yet,

consideratur secundam aliquam particularem causam, necesse est quod in eundem ordinem relabatur secundum aliam causam."—*Sum. Theol.* p. i. q. 19, a. 6, and q. 103, a. 7.

[1] "Unity of Worlds," Essay ii., § ii., p. 260.

[2] See the exceedingly good passage on this subject by the Rev. Dr. Newman, in his "Discourses for Mixed Congregations," 1850, p. 345.

doubtless, not to be compared as to its intensity with the suffering which is produced in man through his high prerogative of self-consciousness.[1]

As to the "beneficial lines" (of Dr. Asa Gray, before referred to), some of the facts noticed in the preceding chapters seem to point very decidedly in that direction, but all must admit that the actual existing outcome is far more "beneficial" than the reverse. The natural universe has resulted in the development of an unmistakable harmony and beauty, and in a decided preponderance of good and of happiness over their opposites.

Even if "laws of nature" did appear, on the theistic hypothesis, to be "superfluous" (which it is by no means intended here to admit), it would be nothing less than puerile to prefer rejecting the hypothesis to conceiving that the appearance of superfluity was probably due to human ignorance; and this especially might be expected from naturalists to whom the interdependence of nature and the harmony and utility of obscure phenomena are becoming continually more clear, as, *e.g.*, the structure of orchids to their illustrious expositor.

Having now cleared the ground somewhat, we may turn to the question what bearing Christian dogma has upon evolution, and whether Christians, as such, need take up any definite attitude concerning it.

As has been said, it is plain that physical science and "evolution" *can* have nothing whatever to do with absolute or primary creation. The Rev. Baden Powell well expresses this, saying : "Science demonstrates incessant past changes, and dimly points to yet earlier links in a more vast series of development of material existence; but the idea of a *beginning*, or of *creation*, in the sense of the original operation of the Divine volition to

---

[1] See Mr. G. H. Lewes's " Sea-Side Studies," for some excellent remarks, beginning at p. 329, as to the small susceptibility of certain animals to pain.

constitute nature and matter, is beyond the province of physical philosophy."[1]

With secondary or derivative creation, physical science is also incapable of conflict; for the objections drawn by some writers seemingly from physical science, are, as has been already argued, rather metaphysical than physical.

Derivative creation is not a supernatural act, but is simply the Divine action by and through natural laws. To recognize such action in such laws is a religious mode of regarding phenomena, which a consistent theist must necessarily accept, and which an atheistic believer must similarly reject. But this conception, if deemed superfluous by any naturalist, can never be shown to be *false* by any investigations concerning natural laws, the constant action of which it presupposes.

The conflict has arisen through a misunderstanding. Some have supposed that by "creation" was necessarily meant either primary, that is, absolute creation, or, at least, some supernatural action; they have therefore opposed the dogma of "creation" in the imagined interest of physical science.

Others have supposed that by "evolution" was necessarily meant a denial of Divine action, a negation of the providence of God. They have therefore combated the theory of "evolution" in the imagined interest of religion.

It appears plain then that Christian thinkers are perfectly free to accept the general evolution theory. But are there any theological authorities to justify this view of the matter?

Now, considering how extremely recent are these biological speculations, it might hardly be expected *a priori* that writers of earlier ages should have given expression to doctrines harmonizing in any degree with such very modern views,[2] nevertheless

[1] "Philosophy of Creation," Essay iii., § iv., p. 480.
[2] It seems almost strange that modern English thought should so long hold aloof from familiar communion with Christian writers of other ages and countries. It is rarely indeed that acquaintance is shown with such

such most certainly is the case, and it would be easy to give numerous examples. It will be better, however, only to cite

authors, though a bright example to the contrary was set by Sir William Hamilton. Sir Charles Lyell (in his "Principles of Geology," 7th edition, p. 35) speaks with approval of the early Italian geologists. Of Vallisneri he says, "I return with pleasure to the geologists of Italy who preceded, as has been already shown, the naturalists of other countries in their investigations into the ancient history of the earth, and who still maintained a decided pre-eminence. They refuted and ridiculed the physico-theological systems of Burnet, Whiston, and Woodward ; while Vallisneri, in his comments on the Woodwardian theory, remarked how much the interests of religion, as well as of those of sound philosophy, had suffered by perpetually mixing up the sacred writings with questions of physical science." Again, he quotes the Carmelite friar Generelli, who, illustrating Moro before the Academy of Cremona in 1749, strongly opposed those who would introduce the supernatural into the domain of nature. " I hold in utter abomination, most learned Academicians ! those systems which are built with their foundations in the air, and cannot be propped up without a miracle, and I undertake, with the assistance of Moro, to explain to you how these marine monsters were transported into the mountains by natural causes."

Sir Charles Lyell notices with exemplary impartiality the spirit of intolerance on both sides. How in France, Buffon, on the one hand, was influenced by the theological faculty of the Sorbonne to recant his theory of the earth, and how Voltaire, on the other, allowed his prejudices to get the better, if not of his judgment, certainly of his expression of it. Thinking that fossil remains of shells, &c., were evidence in favour of orthodox views, Voltaire, Sir Charles Lyell (Principles, p. 56) tells us, " endeavoured to inculcate scepticism as to the real nature of such shells, and to recall from contempt the exploded dogma of the sixteenth century, that they were sports of nature. He also pretended that vegetable impressions were not those of real plants." . . . " He would sometimes, in defiance of all consistency, shift his ground when addressing the vulgar ; and, admitting the true nature of the shells collected in the Alps and other places, pretend that they were Eastern species, which had fallen from the hats of pilgrims coming from Syria. The numerous essays written by him on geological subjects were all calculated to strengthen prejudices, partly because he was ignorant of the real state of the science, and partly from his bad faith." As to the harmony between many early Church writers of great authority and modern views as regards certain matters of geology, see " Geology and Revelation," by the Rev. Gerald Molloy, D.D., London, 1870.

one or two authorities of weight. Now, perhaps no writer of
the earlier Christian ages could be quoted whose authority is
more generally recognized than that of St. Augustin.   The same
may be said of the mediæval period, for St. Thomas Aquinas ; and,
since the movement of Luther, Suarez may be taken as a writer
widely venerated as an authority and one whose orthodoxy has
never been questioned.

It must be borne in mind that for a considerable time after
even the last of these writers no one had disputed the generally
received view as to the small age of the world or at least of the
kinds of animals and plants inhabiting it.   It becomes therefore
much more striking if views formed under such a condition of
opinion are found to harmonize with modern ideas regarding
" Creation " and organic life.

Now St. Augustin insists in a very remarkable manner on the
merely derivative sense in which God's creation of organic forms
is to be understood ; that is, that God created them by conferring
on the material world the power to evolve them under suitable
conditions. He says in his book on Genesis :[1] "Terrestria animalia,
tanquam ex ultimo elemento mundi ultima ; nihilominus *potentialiter*, quorum numeros tempus postea visibiliter explicaret."

Again he says :—

" Sicut autem in ipso grano invisibiliter erant omnia simul,
quæ per tempora in arborem surgerent ; ita ipse mundus cogitan-
dus est, cum Deus *simul omnia creavit*, habuisse simul omnia quæ
in illo et cum illo facta sunt quando factus est dies ; non solum
cœlum cum sole et lunâ et sideribus . . . . ; sed etiam illa quæ
aqua et terra produxit potentialiter atque causaliter, priusquam
per temporum moras ita exorirentur, quomodo nobis jam nota
sunt in eis operibus, quæ Deus usque nunc operatur."[2]

" Omnium quippe rerum quæ corporaliter visibiliterque

[1] " De Genesi ad Litt.," lib. v., cap. v., No. 14 in Ben. Edition,
vol. iii. p. 186.

[2] Lib. cit., cap. xxii., No. 44.

nascuntur, occulta quædam semina in istis corporeis mundi hujus elementis latent." [1]

And again : " Ista quippe originaliter ac primordialiter in quadam textura elementorum cuncta jam creata sunt ; sed acceptis opportunitatibus prodeunt." [2]

St. Thomas Aquinas, as was said in the first chapter, quotes with approval the saying of St. Augustin that in the first institution of nature we do not look for *Miracles*, but for the *laws of Nature* : " In prima institutione naturæ non quæritur miraculum, sed quid natura rerum habeat, ut Augustinus dicit." [3]

Again, he quotes with approval St. Augustin's assertion that the kinds were created only derivatively, "*potentialiter tantum.*" [4]

Also he says, " In prima autem rerum institutione fuit principium activum verbum Dei, quod de materia elementari produxit animalia, vel in actu vel *virtute*, secundum Aug. lib. 5 de Gen. ad lit. c. 5." [5]

Speaking of " kinds " (in scholastic phraseology " substantial forms") latent in matter, he says : " Quas quidam posuerunt non incipere per actionem naturæ sed prius in materia exstitisse, ponentes latitationem formarum. Et hoc accidit eis ex ignorantia materiæ, quia nesciebant distinguere inter potentiam et actum. Quia enim formæ præexistunt eas simpliciter præexistere." [6]

Also Cornelius à Lapide [7] contends that at least certain animals were not absolutely, but only derivatively created, saying of them, " Non fuerunt creata formaliter, sed potentialiter."

As to Suarez, it will be enough to refer to Disp. xv. § 2, n. 9, p. 508, t. i. Edition *Vives*, Paris; also Nos. 13—15, and many

[1] Lib. cit., "De Trinitate," lib. iii., cap. viii, No. 14.
[2] Lib. cit., cap. ix., No. 16.
[3] St. Thomas, Summa, i., quest. 67, art. 4, ad 3.
[4] Primæ Partis, vol. ii., quest. 74, art. 2.
[5] Lib. cit., quest. 71, art. 1.
[6] Lib. cit., quest. 45, art. 8.
[7] *Vide* In Genesim Comment. cap. i.

other references to the same effect could easily be given, but these may suffice.

It is then evident that ancient and most venerable theological authorities distinctly assert *derivative* creation, and thus harmonize with all that modern science can possibly require.

It may indeed truly be said with Roger Bacon, " The saints never condemned many an opinion which the moderns think ought to be condemned." [1]

The various extracts given show clearly how far " evolution " is from any necessary opposition to the most orthodox theology. The same may be said of spontaneous generation. The most recent form of it, lately advocated by Dr. H. Charlton Bastian,[2] teaches that matter exists in two different forms, the crystalline (or statical) and the colloidal (or dynamical) conditions. It also teaches that colloidal matter, when exposed to certain conditions, presents the phenomena of life, and that it can be formed from crystalline matter, and thus that the *prima materia* of which these are diverse forms contains potentially all the multitudinous kinds of animal and vegetable existence. This theory moreover harmonizes well with the views here advocated, for just as crystalline matter builds itself, under suitable conditions, along *certain definite lines*, so analogously colloidal matter has *its definite lines and directions* of development. It is not collected in haphazard, accidental aggregations, but evolves according to its proper laws and special properties.

The perfect orthodoxy of these views is unquestionable. Nothing is plainer from the venerable writers quoted, as well as

---

[1] Roger Bacon, Opus tertium, c. ix. p. 27, quoted in the *Rambler* for 1859, vol. xii. p. 375.

[2] See *Nature*, June and July, 1870. Those who, like Professors Huxley and Tyndall, do not accept his conclusions, none the less agree with him in principle, though they limit the evolution of the organic world from the inorganic to a very remote period of the world's history. (See Professor Huxley's address to the British Association at Liverpool, 1870, p. 17.)

from a mass of other authorities, than that " the supernatural" is not to be looked for or expected in the sphere of mere nature. For this statement there is a general *consensus* of theological authority.

The teaching which the Author has received is, that God is indeed inscrutable and incomprehensible to us from the infinity of His attributes, so that our minds can, as it were, only take in, in a most fragmentary and indistinct manner (as through a glass darkly), dim conceptions of infinitesimal portions of His inconceivable perfection. In this way the partial glimpses obtained by us in different modes differ from each other ; not that God is anything but the most perfect unity, but that apparently conflicting views arise from our inability to apprehend Him, except in this imperfect manner, *i.e.* by successive slight approximations along different lines of approach. Sir William Hamilton has said,[1] "Nature conceals God, and man reveals Him." It is not, according to the teaching spoken of, exactly thus ; but rather that physical nature reveals to us one side, one aspect of the Deity, while the moral and religious worlds bring us in contact with another, and at first, to our apprehension, a very different one. The difference and discrepancy, however, which is at first felt, is soon seen to proceed not from the reason but from a want of flexibility in the imagination. This want is far from surprising. Not only may a man naturally be expected to be an adept in his own art, but at the same time to show an incapacity for a very different mode of activity.[2] We rarely find an artist who takes much interest

[1] "Lectures on Metaphysics and Logic," vol. i. Lecture ii., p. 40.
[2] In the same way that an undue cultivation of any one kind of knowledge is prejudicial to philosophy. Mr. James Martineau well observes, "Nothing is more common than to see maxims, which are unexceptionable as the assumptions of particular sciences, coerced into the service of a universal philosophy, and so turned into instruments of mischief and distortion. That "we can know nothing but phenomena,"— that "causation is simply constant priority,"—that "men are governed invariably by their interests," are examples of rules allowable as dominant hypotheses in physics or political economy, but exercising a desolating tyranny when thrust on

in jurisprudence, or a prizefighter who is an acute metaphysician. Nay, more than this, a positive distaste may grow up, which, in the intellectual order, may amount to a spontaneous and unreasoning disbelief in that which appears to be in opposition to the more familiar concept, and this at all times. It is often and truly said, "that past ages were pre-eminently credulous as compared with our own, yet the difference is not so much in the amount of the credulity, as in the direction which it takes."[1]

Dr. Newman observes : "Any one study, of whatever kind, exclusively pursued, deadens in the mind the interest, nay the perception of any other. Thus Cicero says, that Plato and Demosthenes, Aristotle and Isocrates, might have respectively excelled in each other's province, but that each was absorbed in his own. Specimens of this peculiarity occur every day. You can hardly persuade some men to talk about anything but their own pursuit ; they refer the whole world to their own centre, and measure all matters by their own rule, like the fisherman in the drama, whose eulogy on his deceased lord was 'he was so fond of fish.'"[2]

The same author further says :[3] "When anything, which comes before us, is very unlike what we commonly experience, we consider it on that account untrue ; not because it really shocks our reason as improbable, but because it startles our imagination as strange. Now, revelation presents to us a perfectly different aspect of the universe from that presented by the sciences. The two informations are like the distinct subjects represented by the lines of the same drawing, which, accordingly

to the throne of universal empire. He who seizes upon these and similar maxims, and carries them in triumph on his banner, may boast of his escape from the uncertainties of metaphysics, but is himself all the while the unconscious victim of their very vulgarest deception." ("Essays," Second Series, *A Plea for Philosophical Studies,* p. 421.)

[1] Lecky's "History of Rationalism," vol. i. p. 73.
[2] "Lectures on University Subjects," by J. H. Newman, D.D., p. 322.
[3] Loc. cit. p. 324.

as they are read on their concave or convex side, exhibit to us now a group of trees with branches and leaves, and now human faces." . . . " While then reason and revelation are consistent in fact, they often are inconsistent in appearance ; and this seeming discordance acts most keenly on the imagination, and may suddenly expose a man to the temptation, and even hurry him on to the commission of definite acts of unbelief, in which reason itself really does not come into exercise at all." [1]

Thus we find in fact just that distinctness between the ideas derived from physical science on the one hand and from religion on the other, which we might *a priori* expect if there exists that distinctness between the natural and the miraculous which theological authorities lay down.

Assuming, for argument's sake, the truth of Christianity, it evidently has not been the intention of its Author to make the evidence for it so plain that its rejection would be the mark of intellectual incapacity. Conviction is not forced upon men in the way that the knowledge that the government of England is constitutional, or that Paris is the capital of France, is forced upon all who choose to inquire into those subjects. The Christian system is one which puts on the strain, as it were, *every* faculty of man's nature, and the intellect is not (any more than we should *a priori* expect it to be) exempted from taking part in the probationary trial. A moral element enters into the acceptance of that system.

And so with natural religion—with those ideas of the supernatural, viz. God, Creation, and Morality, which are anterior to revelation and repose upon reason. Here again it evidently has not been the intention of the Creator to make the evidence of His existence so plain that its non-recognition would be

[1] Thus Professor Tyndall, in the *Pall Mall Gazette* of June 15, 1868, speaking of physical science, observes, " The *logical feebleness* of science is not sufficiently borne in mind. It keeps down the weed of superstition, not by logic, but by slowly rendering the mental soil unfit for its cultivation."

the mark of intellectual incapacity. Conviction, as to theism, is not forced upon men as is the conviction of the existence of the sun at noon-day.[1] A moral element enters also here, and the analogy there is in this respect between Christianity and theism speaks eloquently of their primary derivation from one common author.

Thus we might expect that it would be a vain task to seek anywhere in nature for evidence of Divine action, such that no one could sanely deny it. God will not allow Himself to be caught at the bottom of any man's crucible, or yield Himself to the experiments of gross-minded and irreverent inquirers. The natural, like the supernatural, revelation appeals to *the whole* of man's mental nature and not to the *reason alone.*[2]

None, therefore, need feel disappointed that evidence of the direct action of the first cause in merely natural phenomena ever eludes our grasp ; for assuredly those same phenomena will ever remain fundamentally inexplicable by physical science alone.

There being then nothing in either authority or reason which makes "evolution" repugnant to Christianity, is there anything in the Christian doctrine of "Creation" which is repugnant to the theory of "evolution"?

Enough has been said as to the distinction between absolute and derivative "creation." It remains to consider the successive "evolution" (Darwinian and other) of "specific forms," in a theological light.

As to what "evolution" is, we cannot of course hope to explain it completely, but it may be enough to define it as the manifestation to the intellect, by means of sensible impressions, of some ideal entity (power, principle, nature, or activity)

---

[1] By this it is not, of course, meant to deny that the existence of God can be demonstrated so as to demand the assent of the intellect taken, so to speak, by itself.

[2] See some excellent remarks in the Rev. Dr. Newman's Parochial Sermons—the new edition (1869), vol. i. p. 211.

which before that manifestation was in a latent, unrealized, and merely "potential" state—a state that is capable of becoming realized, actual, or manifest, the requisite conditions being supplied.

"Specific forms," kinds or species, are (as was said in the introductory chapter) "peculiar congeries of characters or attributes, innate powers and qualities, and a certain nature realized in individuals."

Thus, then, the "evolution of specific forms" means the actual manifestation of special powers, or natures, which before were latent, in such a successive manner that there is in some way a genetic relation between posterior manifestations and those which preceded them.

On the special Darwinian hypothesis the manifestation of these forms is determined simply by the survival of the fittest of many indefinite variations.

On the hypothesis here advocated the manifestation is controlled and helped by such survival, but depends on some unknown internal law or laws which determine variation at special times and in special directions.

Professor Agassiz objects to the evolution theory, on the ground that "species, genera, families, &c., exist as thoughts, individuals as facts,"[1] and he offers the dilemma, "If species do not exist at all, as the supporters of the transmutation theory maintain, how can they vary? and if individuals alone exist, how can the differences which may be observed among them prove the variability of species?"

But the supporter of "evolution" need only maintain that the several "kinds" become manifested gradually by slight differences among the various individual embodiments of one specific idea. He might reply to the dilemma by saying, species do not exist *as species* in the sense in which they are said to vary (variation applying only to the concrete embodiments of

---

[1] *American Journal of Science*, July 1860, p. 143, quoted in Dr. Asa Gray's pamphlet, p. 47.

the specific idea), and the evolution of species is demonstrated not by individuals *as individuals*, but as embodiments of different specific ideas.

Some persons seem to object to the term "creation" being applied to evolution, because evolution is an "exceedingly slow and gradual process." Now even if it were demonstrated that such is really the case, it may be asked, what is "slow and gradual"? The terms are simply relative, and the evolution of a specific form in ten thousand years would be instantaneous to a being whose days were as hundreds of millions of years.

There are others again who are inclined absolutely to deny the existence of species altogether, on the ground that their evolution is so gradual that if we could see all the stages it would be impossible to say *when* the manifestation of the old specific form ceased and that of the new one began. But surely it is no approach to a reason against the existence of a thing that we cannot determine the exact moment of its first manifestation. When watching "dissolving views," who can tell, whilst closely observing the gradual changes, exactly at what moment a new picture, say St. Mark's, Venice, can be said to have commenced its manifestation, or have begun to dominate a preceding representation of "Dotheboys' Hall"? That, however, is no reason for denying the complete difference between the two pictures and the ideas they respectively embody.

The notion of a special nature, a peculiar innate power and activity—what the scholastics called a "substantial form"—will be distasteful to many. The objection to the notion seems, however, to be a futile one, for it is absolutely impossible to altogether avoid such a conception and such an assumption. If we refuse it to the individuals which embody the species, we must admit it as regards their component parts—nay, even if we accept the hypothesis of pangenesis, we are nevertheless compelled to attribute to each gemmule that peculiar power of reproducing its own nature (its own "substantial form"), with its special activity, and that remarkable power of annexing itself to certain other well-defined gemmules whose nature it is also to plant themselves in a certain definite vicinity. So that in each individual, instead of one such peculiar power and activity dominating and controlling all the parts, you have an infinity of separate powers and activities limited to the several minute component gemmules.

It is possible that in some minds, the notion may lurk that such powers are simpler and easier to understand, because the bodies they affect are so minute! This absurdity hardly bears stating. We can easily conceive a being so small, that a gemmule would be to it as large as St. Paul's would be to us.

Admitting then the existence of species, and of their successive evolution, is there anything in these ideas hostile to Christian belief?

Writers such as Vogt and Buchner will of course contend that there is; but naturalists, generally, assume that God acts in and by the various laws of nature. And this is equivalent to admitting the doctrine of "derivative creation." With very few exceptions, none deny such Divine concurrence. Even "design" and "purpose" are recognized as quite compatible with evolution, and even with the special "nebular" and Darwinian forms of it. Professor Huxley well says,[1] "It is necessary to remark that there is a wider teleology, which is not touched by the doctrine of evolution, but is actually based upon the fundamental proposition of evolution." . . . . "The teleological and the mechanical views of nature are not necessarily mutually exclusive; on the contrary, the more purely a mechanist the speculator is, the more firmly does he assume a primordial molecular arrangement, of which all the phenomena of the universe are the consequences; and the more completely thereby is he at the mercy of the teleologist, who can always defy him to

[1] See *The Academy* for October 1869, No. 1, p. 13.

disprove that this primordial molecular arrangement was not intended to evolve the phenomena of the universe."[1]

Professor Owen says, that natural evolution, through secondary causes, " by means of slow physical and organic operations through long ages, is not the less clearly recognizable as the act of all adaptive mind, because we have abandoned the old error of supposing it to be the result[2] of a primary, direct, and sudden act of creational construction." . . . "The succession of species by continuously operating law, is not necessarily a ' blind operation.' Such law, however discerned in the properties and successions of natural objects, intimates, nevertheless, a preconceived progress. Organisms may be evolved in orderly succession, stage after stage, towards a foreseen goal, and the broad features of the course may still show the unmistakable impress of Divine volition."

Mr. Wallace[3] declares that the opponents of evolution present a less elevated view of the Almighty. He says : " Why should we suppose the machine too complicated to have been designed by the Creator so complete that it would necessarily work out harmonious results ? The theory of ' continual interference' is a limitation of the Creator's power. It assumes that He could not work by pure law in the organic, as He has done in the inorganic world." Thus, then, there is not only no necessary antagonism between the general theory of " evolution " and a Divine action, but the compatibility between the two is recognized by naturalists who cannot be suspected of any strong theological bias.

[1] Professor Huxley goes on to say that the mechanist may, in turn, demand of the teleologist how the latter knows it was so intended. To this it may be replied he knows it as a necessary truth of reason deduced from his own primary intuitions, which intuitions cannot be questioned without *absolute* scepticism.

[2] The Professor doubtless means the *direct* and *immediate* result. (See Trans. Zool. Soc. vol. v. p. 90.)

[3] " Natural Selection," p. 280.

The very same may be said as to the special Darwinian form of the theory of evolution.

It is true Mr. Darwin writes sometimes as if he thought that his theory militated against even *derivative creation.*[1] This, however, there is no doubt, was not really meant; and indeed, in the passage before quoted and criticised, the possibility of the Divine ordination of each variation is spoken of as a tenable view. He says ("Origin of Species," p. 569), "I see no good reason why the views given in this volume should shock the religious feelings of any one;" and he speaks of life "having been originally breathed by the Creator into a few forms or into one," which is *more* than the dogma of creation actually requires.   We find then that no *in*compatibility is asserted (by any scientific writers worthy of mention) between "evolution" and the co-operation of the Divine will; while the same "evolution" has been shown to be thoroughly acceptable to the most orthodox theologians who repudiate the intrusion of the supernatural into the domain of nature.   A more complete harmony could scarcely be desired.

But if we may never hope to find, in physical nature, evidence of supernatural action, what sort of action might we expect to find there, looking at it from a theistic point of view? Surely an action the results of which harmonize with man's reason,[2] which

---

[1] Dr. Asa Gray, *e.g.*, has thus understood Mr. Darwin. The Doctor says in his pamphlet, p. 38, "Mr. Darwin uses expressions which imply that the natural forms which surround us, because they have a history or natural sequence, could have been only generally, but not particularly designed,—a view at once superficial and contradictory; whereas his true line should be, that his hypothesis concerns the *order* and not the *cause*, the *how* and not the *why* of the phenomena, and so leaves the question of design just where it was before."

[2] "All science is but the partial reflexion in the *reason of man*, of the great all-pervading *reason of the universe*. And the *unity* of science is the reflexion of the *unity* of nature and of the *unity* of that supreme reason and intelligence which pervades and rules over nature, and from whence all reason and all science is derived." (Rev. Baden Powell, "Unity of the Sciences," Essay i. § ii. p. 81.)

is orderly, which disaccords with the action of blind chance and with the "fortuitous concourse of atoms" of Democritus; but at the same time an action which, as to its modes, ever, in parts, and in ultimate analysis, eludes our grasp, and the modes of which are different from those by which we should have attempted to accomplish such ends.

Now, this is just what we *do* find. The harmony, the beauty, and the order of the physical universe are the themes of continual panegyrics on the part of naturalists, and Mr. Darwin, as the Duke of Argyll remarks,[1] " exhausts every form of words and of illustration by which intention or mental purpose can be described "[2] when speaking of the wonderfully complex adjustments to secure the fertilization of orchids. Also, we find co-existing with this harmony a mode of proceeding so different from that of man as (the direct supernatural action eluding us) to form a stumbling-block to many in the way of their recognition of Divine action at all : although nothing can be more inconsistent than to speak of the first cause as utterly inscrutable and incomprehensible, and at the same time to expect to find traces of a mode of action exactly similar to our own. It is surely enough if the results harmonize on the whole and preponderatingly with the rational, moral, and æsthetic instincts of man.

Mr. J. J. Murphy [3] has brought strongly forward the evidence of "intelligence" throughout organic nature. He believes "that there is something in organic progress which mere Natural Selection among spontaneous variations will not account for," and that "this something is that organizing intelligence which guides the action of the inorganic forces, and forms structures which neither Natural Selection nor any other unintelligent agency could form."

---

[1] " The Reign of Law," p. 40.

[2] Though Mr. Darwin's epithets denoting design are metaphorical, his admiration of the result is unequivocal, nay, enthusiastic !

[3] See "Habit and Intelligence," vol. i. p. 348.

This intelligence, however, Mr. Murphy considers may be unconscious, a conception which it is exceedingly difficult to understand, and which to many minds appears to be little less than a contradiction in terms ; the very first condition of an intelligence being that, if it knows anything, it should at least know its own existence.

Surely the evidence from physical facts agrees well with the overruling, concurrent action of God in the order of nature; which is no miraculous action, but the operation of laws which owe their foundation, institution, and maintenance to an omniscient Creator of whose intelligence our own is a feeble adumbration, inasmuch as it is created in the "image and likeness" of its Maker.

This leads to the final consideration, a difficulty by no means to be passed over in silence, namely the ORIGIN OF MAN. To the general theory of Evolution, and to the special Darwinian form of it, no exception, it has been shown, need be taken on the ground of orthodoxy. But in saying this, it has not been meant to include the soul of man.

It is a generally received doctrine that the soul of every individual man is absolutely created in the strict and primary sense of the word, that it is produced by a direct or supernatural[1] act, and, of course, that by such an act the soul of the first man was similarly created. It is therefore important to inquire whether "evolution" conflicts with this doctrine.

Now the two beliefs are in fact perfectly compatible, and that either on the hypothesis—1. That man's body was created in a manner different in kind from that by which the bodies of other animals were created ; or 2. That it was created in a similar manner to theirs.

One of the authors of the Darwinian theory, indeed, contends

---

[1] The term, as before said, not being used in its ordinary theological sense, but to denote an immediate Divine action as distinguished from God's action through the powers conferred on the physical universe.

that even as regards man's body, an action took place different from that by which brute forms were evolved. Mr. Wallace [1] considers that "Natural Selection" alone could not have produced so large a brain in the savage, in possessing which he is furnished with an organ beyond his needs. Also that it could not have produced that peculiar distribution of hair, especially the nakedness of the back, which is common to all races of men, nor the peculiar construction of the feet and hands. He says,[2] after speaking of the prehensile foot, common without a single exception to all the apes and lemurs, "It is difficult to see why the prehensile power should have been taken away" by the mere operation of Natural Selection. "It must certainly have been useful in climbing, and the case of the baboons shows that it is quite compatible with terrestrial locomotion. It may not be compatible with perfectly easy erect locomotion; but, then, how can we conceive that early man, *as an animal*, gained anything by purely erect locomotion? Again, the hand of man contains latent capacities and powers which are unused by savages, and must have been even less used by palæolithic man and his still ruder predecessors. It has all the appearance of an organ prepared for the use of civilized man, and one which was required to render civilization possible." Again speaking of the "wonderful power, range, flexibility, and sweetness of the musical sounds producible by the human larynx," he adds, "The habits of savages give no indication of how this faculty could have been developed by Natural Selection; because it is never required or used by them. The singing of savages is a more or less monotonous howling, and the females seldom sing at all. Savages certainly never choose their wives for fine voices, but for rude health, and strength and physical beauty. Sexual selection could not therefore have developed this wonderful power, which only comes into play among civilized people. It seems as if the organ had been prepared in anticipation of the future

[1] See "Natural Selection," pp. 332 to 360.　　　[2] Loc. cit., p. 349.

progress of man, since it contains latent capacities which are useless to him in his earlier condition. The delicate correlations of structure that give it such marvellous powers, could not therefore have been acquired by means of Natural Selection."

To this may be added the no less wonderful faculty in the ear of appreciating delicate musical tones, and the harmony of chords.

It matters not what part of the organ subserves this function, but it has been supposed that it is ministered to by the fibres *of Corti.*[1] Now it can hardly be contended that the preservation

FIBRES OF CORTI.

of any race of men in the struggle for life could have depended on such an extreme delicacy and refinement of the internal ear,[2]—a perfection only fully exercised in the enjoyment and appreciation of the most exquisite musical performances. Here, surely, we have an instance of an organ preformed, ready beforehand for such action as could never by itself have been the cause of its development,—the action having only been subsequent, not anterior. The Author is not aware what may be the minute structure of the internal ear in the highest apes, but if (as from analogy is probable) it is much as in man, then *a fortiori* we have an instance of *anticipatory* development of a most

[1] See Professor Huxley's "Lessons in Elementary Physiology," p. 218.

[2] It may be objected, perhaps, that excessive delicacy of the ear might have been produced by having to guard against the approach of enemies, some savages being remarkable for their keenness of hearing at great distances. But the perceptions of *intensity* and *quality* of sound are very different. Some persons who have an extremely acute ear for delicate sounds, and who are fond of music, have yet an incapacity for detecting whether an instrument is slightly out of tune.

marked and unmistakable kind. And this is not all. There
is no reason to suppose that any animal besides man appreciates
musical *harmony*. It is certain that no other one *produces* it.

Mr. Wallace also urges objections drawn from the origin of
some of man's mental faculties, such as "the capacity to form
ideal conceptions of space and time, of eternity and infinity—
the capacity for intense artistic feelings of pleasure, in form,
colour and composition—and for those abstract notions of form
and number which render geometry and arithmetic possible,"
also from the origin of the moral sense.[1]

The validity of these objections is fully conceded by the
Author of this book, but he would push it much further,
and contend (as has been now repeatedly said), that another
law, or other laws, than "Natural Selection" have determined
the evolution of *all* organic forms, and of inorganic forms also.
And it must be contended that Mr. Wallace, in order to be quite
self-consistent, should arrive at the very same conclusion, inasmuch as he is inclined to trace all phenomena to the action of
superhuman WILL. He says : [2] "If therefore we have traced one
force, however minute, to an origin in our own WILL, while we
have no knowledge of any other primary cause of force, it does
not seem an improbable conclusion that all force may be will-
force ; and thus, that the whole universe is not merely dependent
on, but actually *is*, the WILL of higher intelligences, or of one
Supreme Intelligence."

If there is really evidence, as Mr. Wallace believes, of the
action of an overruling intelligence in the evolution of the
"human form divine ;" if we may go so far as this, then surely
an analogous action may well be traced in the production of the
horse, the camel, or the dog, so largely identified with human
wants and requirements. And if from other than physical considerations we may believe that such action, though undemon-
strable, has been and is ; then (reflecting on sensible phenomena

[1] Loc. cit., pp. 351, 352.   [2] Loc. cit., p. 368.

the theistic light derived from psychical facts) we may, in the language of Mr. Wallace, " see indications of that power in facts which, by themselves, would not serve to prove its existence." [1]

Mr. Murphy, as has been said before, finds it necessary to accept the wide-spread action of " intelligence " as the agent by which *all* organic forms have been called forth from the inorganic. But all science tends to unity, and this tendency makes it reasonable to extend to all physical existences a mode of formation which we may have evidence for in any *one* of them. It therefore makes it reasonable to extend, if possible, the very same agency which we find operating in the field of biology, also to the inorganic world. If on the grounds brought forward the action of intelligence may be affirmed in the production of man's bodily structure, it becomes probable *a priori* that it may also be predicated of the formative action by which has been produced the animals which minister to him, and all organic life whatsoever. Nay more, it is then congruous to expect analogous action in the development of crystalline and colloidal structures, and in that of all chemical compositions, in geological evolutions, and the formation not only of this earth, but of the solar system and whole sidereal universe.

If such really be the direction in which physical science, philosophically considered, points ; if intelligence may thus be seen to preside over the evolution of each system of worlds and the unfolding of every blade of grass—this grand result harmonizes indeed with the teachings of faith that God acts and concurs, in the natural order, with those laws of the material universe which were not only instituted by His will, but are sustained by His concurrence ; and we are thus enabled to discern in the natural order, however darkly, the Divine Author of nature—Him in whom " we live, and move, and have our being."

But if this view is accepted, then it is no longer absolutely

[1] Loc. cit., p. 350.

necessary to suppose that any action different in kind took place in the production of man's body, from that which took place in the production of the bodies of other animals, and of the whole material universe.

Of course, if it *can* be demonstrated that that difference which Mr. Wallace asserts really exists, it is plain that we then have to do with facts not only harmonizing with religion, but, as it were, preaching and proclaiming it.

It is not, however, necessary for Christianity that any such view should prevail. Man, according to the old scholastic definition, is " a rational animal" (*animal rationale*), and his animality is distinct in nature from his rationality, though inseparably joined, during life, in one common personality. This animal body must have had a different source from that of the spiritual soul which informs it, from the distinctness of the two orders to which those two existences severally belong.

Scripture seems plainly to indicate this when it says that " God made man from the dust of the earth, and breathed into his nostrils the breath of life." This is a plain and direct statement that man's *body* was *not* created in the primary and absolute sense of the word, but was evolved from pre-existing material (symbolized by the term " dust of the earth "), and was therefore only *derivatively created*, i.e. by the operation of secondary laws. His *soul*, on the other hand, was created in quite a different way, not by any pre-existing means, external to God himself, but by the direct action of the Almighty, symbolized by the term " breathing : " the very form adopted by Christ, when conferring the *supernatural* powers and graces of the Christian dispensation, and a form still daily used in the rites and ceremonies of the Church.

That the first man should have had this double origin agrees with what we now experience. For supposing each human soul to be directly and immediately created, yet each human body is evolved by the ordinary operation of natural physical laws.

Professor Flower in his Introductory Lecture[1] (p. 20) to his course of Hunterian Lectures for 1870 well observes: "Whatever man's place may be, either *in* or *out* of nature, whatever hopes, or fears or feelings about himself or his race he may have, we all of us admit that these are quite uninfluenced by our knowledge of the fact that each individual man comes into the world by the ordinary processes of generation, according to the same laws which apply to the development of all organic beings whatever, that every part of him which can come under the scrutiny of the anatomist or naturalist, has been evolved according to these regular laws from a simple minute ovum, indistinguishable to our senses from that of any of the inferior animals. If this be so—if man is what he is, notwithstanding the corporeal mode of origin of the individual man, so he will assuredly be neither less nor more than man, whatever may be shown regarding the corporeal origin of the whole race, whether this was from the dust of the earth, or by the modification of some pre-existing animal form."

Man is indeed compound, in him two distinct orders of being impinge and mingle ; and with this an origin from two concurrent modes of action is congruous, and might be expected *a priori.* At the same time as the " soul " is " the form of the body,'' the former might be expected to modify the latter into a structure of harmony and beauty standing alone in the organic world of nature. Also that, with the full perfection and beauty of that soul, attained by the concurrent action of " Nature " and " Grace," a character would be formed like nothing else which is visible in this world, and having a mode of action different, inasmuch as complementary to all inferior modes of action.

Something of this is evident even to those who approach the subject from the point of view of physical science only. Thus Mr. Wallace observes,[2] that on his view man is to be placed

---

[1] Published by John Churchill.   [2] Natural Selection, p. 324.

"apart, as not only the head and culminating point of the grand series of organic nature, but as in some degree *a new and distinct order of being.*[1] From those infinitely remote ages when the first rudiments of organic life appeared upon the earth, every plant and every animal has been subject to one great law of physical change. As the earth has gone through its grand cycles of geological, climatal, and organic progress, every form of life has been subject to its irresistible action, and has been continually but imperceptibly moulded into such new shapes as would preserve their harmony with the ever-changing universe. No living thing could escape this law of its being; none (except, perhaps, the simplest and most rudimentary organisms) could remain unchanged and live amid the universal change around it."

"At length, however, there came into existence a being in whom that subtle force we term *mind*, became of greater importance than his mere bodily structure. Though with a naked and unprotected body, *this* gave him clothing against the varying inclemencies of the seasons. Though unable to compete with the deer in swiftness, or with the wild bull in strength, *this* gave him weapons with which to capture or overcome both. Though less capable than most other animals of living on the herbs and the fruits that unaided nature supplies, this wonderful faculty taught him to govern and direct nature to his own benefit, and make her produce food for him when and where he pleased. From the moment when the first skin was used as a covering; when the first rude spear was formed to assist in the chase; when fire was first used to cook his food; when the first seed was sown or shoot planted, a grand revolution was effected in nature, a revolution which in all the previous ages of the earth's history had had no parallel, for a being had arisen who was no longer necessarily subject to change with the changing universe, a being who was in some degree superior to nature, inasmuch as he knew how to control and regulate her action, and could keep

[1] The italics are not Mr. Wallace's.

himself in harmony with her, not by a change in body, but by an advance in mind."

"On this view of his special attributes, we may admit 'that he is indeed a being apart.' Man has not only escaped 'Natural Selection' himself, but he is actually able to take away some of that power from nature which before his appearance she universally exercised. We can anticipate the time when the earth will produce only cultivated plants and domestic animals; when man's selection shall have supplanted 'Natural Selection;' and when the ocean will be the only domain in which that power can be exerted."

Baden Powell[1] observes on this subject: "The relation of the animal man to the intellectual, moral, and spiritual man, resembles that of a crystal slumbering in its native quarry to the same crystal mounted in the polarizing apparatus of the philosopher. The difference is not in physical nature, but in investing that nature with a new and higher application. Its continuity with the material world remains the same, but a new relation is developed in it, and it claims kindred with ethereal matter and with celestial light."

This well expresses the distinction between the merely physical and the hyperphysical natures of man, and the subsumption of the former into the latter which dominates it.

The same author in speaking of man's moral and spiritual nature says,[2] "The assertion in its very nature and essence refers wholly to a DIFFERENT ORDER OF THINGS, apart from and transcending any material ideas whatsoever." Again[3] he adds, "In proportion as man's *moral* superiority is held to consist in attributes *not* of a *material* or corporeal kind or origin, it can signify little how his *physical* nature may have originated."

Now physical science, as such, has nothing to do with the soul of man which is hyperphysical. That such an entity exists,

---

[1] "Unity of Worlds," Essay ii. § ii. p. 247.
[2] Ibid. Essay i. § ii. p. 76.   [3] Ibid. Essay iii. § iv. p. 466.

that the correlated physical forces go through their Protean
transformations, have their persistent ebb and flow outside of
the world of WILL and SELF-CONSCIOUS MORAL BEING, are propositions the proofs of which have no place in this work. This
at least may however be confidently affirmed, that no reach of
physical science in any coming century will ever approach to a
demonstration that countless modes of being, as different from
each other as are the force of gravitation and conscious maternal
love, may not co-exist. Two such modes are made known to us
by our natural faculties only : the physical, which includes the
first of these examples ; the hyperphysical, which embraces the
other. For those who accept revelation, a third and a distinct
mode of being and of action is also made known, namely, the
direct and immediate or, in the sense here given to the term,
the supernatural. An analogous relationship runs through and
connects all these modes of being and of action. The higher
mode in each case employs and makes use of the lower, the
action of which it occasionally suspends or alters, as gravity is
suspended by electro-magnetic action, or the living energy of
an organic being restrains the inter-actions of the chemical
affinities belonging to its various constituents.

Thus conscious will controls and directs the exercise of the
vital functions according to desire, and moral consciousness tends
to control desire in obedience to higher dictates.[1] The action of

[1] A good exposition of how an inferior action has to yield to one higher
is given by Dr. Newman in his "Lectures on University Subjects," p. 372.
" What is true in one science, is dictated to us indeed according to that
science, but not according to another science, or in another department.
" What is certain in the military art, has force in the military art, but
not in statesmanship ; and if statesmanship be a higher department of
action than war, and enjoins the contrary, it has no force on our reception
and obedience at all. And so what is true in medical science, might in all
cases be carried out, *were* man a mere animal or brute without a soul ;
but since he is a rational, responsible being, a thing may be ever so true in
medicine, yet may be unlawful in fact, in consequence of the *higher* law of
morals and religion coming to some different conclusion."

living organisms depends upon and subsumes the laws of inorganic matter. Similarly the actions of animal life depend upon and subsume the laws of organic matter. In the same way the actions of a self-conscious moral agent, such as man, depend upon and subsume the laws of animal life. When a part or the whole series of these natural actions is altered or suspended by the intervention of action of a still higher order, we have then a " miracle."

In this way we find a perfect harmony in the double nature of man, his rationality making use of and subsuming his animality; his soul arising from direct and immediate creation, and his body being formed at first (as now in each separate individual) by derivative or secondary creation, through natural laws. By such secondary creation, *i.e.* by natural laws, for the most part as yet unknown but controlled by " Natural Selection," all the various kinds of animals and plants have been manifested on this planet. That Divine action has concurred and concurs in these laws we know by deductions from our primary intuitions ; and physical science, if unable to demonstrate such action, is at least as impotent to disprove it. Disjoined from these deductions, the phenomena of the universe present an aspect devoid of all that appeals to the loftiest aspirations of man, that which stimulates his efforts after goodness, and presents consolations for unavoidable shortcomings. Conjoined with these same deductions, all the harmony of physical nature and the constancy of its laws are preserved unimpaired, while the reason, the conscience, and the æsthetic instincts are alike gratified. We have thus a true reconciliation of science and religion, in which each gains and neither loses, one being complementary to the other.

Some apology is due to the reader for certain observations and arguments which have been here advanced, and which have little in the shape of novelty to recommend them. But after all, novelty can hardly be predicated of the views here criticised and opposed. Some of these seem almost a return to the "fortuitous concourse of atoms" of Democritus, and even the very theory of "Natural Selection" itself—a "survival of the fittest" —was in part thought out not hundreds but *thousands* of years ago. Opponents of Aristotle maintained that by the accidental occurrence of combinations, organisms have been preserved and perpetuated such as final causes, did they exist, would have brought about, disadvantageous combinations or variations being speedily exterminated. "For when the very same combinations happened to be produced which the law of final causes would have called into being, those combinations which proved to be advantageous to the organism were preserved; while those which were not advantageous perished, and still perished like the minotaurs and sphinxes of Empedocles."[1]

In conclusion, the Author ventures to hope that this treatise may not be deemed useless, but have contributed, however slightly, towards clearing the way for peace and conciliation and for a more ready perception of the harmony which exists between those deductions from our primary intuitions before alluded to, and the teachings of physical science, as far, that is, as concerns the evolution of organic forms—*the genesis of species.*

The aim has been to support the doctrine that these species have been evolved by ordinary *natural laws* (for the most part unknown) controlled by the *subordinate* action of "Natural Selection," and at the same time to remind some that there is and can be absolutely nothing in physical science which forbids them to regard those natural laws as acting with the Divine concurrence and in obedience to a creative fiat originally imposed on the primeval Cosmos, "in the beginning," by its Creator, its Upholder, and its Lord.

---

[1] Quoted from the *Rambler* of March 1860, p. 364: "῝Οπου μὲν οὖν ἅπαντα συνέβη, ὥσπερ κἂν εἰ ἕνεκά του ἐγίνετο, ταῦτα μὲν ἐσώθη ἀπὸ τοῦ αὐτομάτου συστάντα ἐπιτηδείως, ὅσα δὲ μὴ οὕτως ἀπώλετο καὶ ἀπόλλυται, καθάπεο Ἐμπεδοκλῆς λέγει τὰ βουγενῆ καὶ ἀνδρίπρωρα."—ARIST. *Phys.* ii. c. 8.

INDEX.

A.

AARD-VARK, 174.
Absolute creation, 252.
Acanthometræ, 186.
Acrodont teeth, 148.
Acts formally moral, 195.
Acts materially moral, 195.
Adductor muscles, 79.
Agassiz, Professor, 271.
Aged, care of, 192.
Aggregational theory, 163.
Algoa Bay, cat of, 98.
Allantois, 82.
Amazons, butterflies of, 85.
Amazons, cholera in the, 192.
American butterflies, 29.
American maize, 100.
American monkeys, 226.
Amiurus, 147.
Amphibia, 109.
Analogical relations, 157.
Ancon sheep, 100, 103, 227.
Andrew Murray, Mr., 83.
Angora cats, 175.
Animal's sufferings, 260.
Ankle bones, 158.
Annelids undergoing fission, 169, 211.
Annulosa, eye of, 76.
Anoplotherium, 109.
Anteater, 83.
Antechinus, 82.
Antenna, of orchid, 56.
Anthropomorphism, 258.
Ape's sexual characters, 49.
Apostles' Creed, 245.
Appendages of lobster, 161.
Appendages of Normandy pigs, 99.
Appendages of turkey, 100.
Appendix, vermiform, 83.

Appreciation of Mr. Darwin, 10.
Apteryx, 7, 70.
Aqueous humour, 76.
Aquinas, St. Thomas, 17, 263, 265.
Archegosaurus, 135.
Archeopteryx, 73.
Arcturus, 193.
Argyll, Duke of, 14, 276.
Aristotle, 288.
Armadillo, extinct kind, 110.
Arthritis, rheumatic, 183.
Artiodactyle foot, 109.
Asa Gray, Dr., 253, 255, 261.
Asceticism, 193.
Ascidians, placental structure, 81.
Assumptions of Mr. Darwin, 16.
Astronomical objections, 136.
Auditory organ, 74.
Augustin, St., 17, 263, 264.
Aurelius, Marcus, 206.
Avian limb, 106.
Avicularia, 80.
Axolotl, 165.
Aye-Aye, 107.
Aylesbury ducks, 234.

B.

BACKBONE, 135, 162.
Bacon, Roger, 266.
Baleen, 40.
Bamboo insect, 33.
Bandicoot, 67.
Bartlett, Mr. A. D., 126, 234.
Bartlett, Mr. E., 192.
Basil, St., 17.
Bastian, Dr. H. Charlton, 115, 219 237, 266.
Bat, wing of, 64.
Bates, Mr., 29, 85, 87.

U

Bats, 108.
Beaks, 83.
Beasts, sufferings of, 260.
Beauty of shell-fish, 54.
Bee orchid, 55.
Bird, wings of, 64.
Birds compared with reptiles, 70.
Bird's-head processes, 80.
Birds of Paradise, 90.
Birth of individual and species, 2.
Bivalves, 79.
Black sheep, 122.
Black-shouldered peacock, 100.
Bladebone, 70.
Blood-vessels, 182.
Blyth, Mr., 100, 181.
Bones of skull, 153.
Bonnet, M., 217.
Borwick, Mr., 198.
"Boots" of pigeons, 181.
Breathing, modified power of, 99.
Breeding of lions, 234.
Brill, 37.
Broccoli, variety of, 100.
Bryozoa, 81.
Buchner, Dr., 273.
Budd, Dr. W., 183.
Buffon, 217.
Bull-dog's instinct, 260.
Burt, Prof. Wilder, 180, 184.
Butterflies, 29.
Butterflies, Amazonian, 85.
Butterflies, American, 29.
Butterflies of Indian region, 83.
Butterflies, tails of, 85.
Butterfly, Leaf, 31.

C.

Cacotus, 149.
Cæcum, 83.
Calamaries, 77.
Cambrian deposits, 137.
Cape ant-eater, 174.
Care of aged, 192.
Carinate birds, 70.
Carnivora, 68.
Carnivorous dentition, 110.
Carp fishes, 146.
Carpal bones, 106, 178.
Carpenter, Dr., 115.
Carpus, 177, 178.
Cases of conscience, 201.
Cassowary, 70.

Catasetum, 56.
Causes of spread of Darwinism, 10.
Cebus, 226.
Celebes, butterflies of, 85.
Centetes, 148.
Centipede, 66, 159.
Cephalopoda, 74.
Ceroxylus laceratus, 36.
Cetacea, 42, 83, 105, 108, 174.
Chances against few individuals, 57.
Characinidæ, 146.
Cheirogaleus, 158.
Chetahs, 234.
Chickens, mortality of hybrids, 124.
Chioglossa, 165.
Chiromys, 107.
Cholera, 192.
Choroid, 76.
Chronic rheumatism, 183.
Circumcision, 212.
Clarias, 146.
Climate, effects of, 98.
Climbing plants, 107.
Clock-thinking illustration, 249.
Cobra, 50.
Cockle, 79.
Cod, 39.
Colloidal matter, 266.
Conceptions, symbolic, 251.
Connecticut footsteps, 131.
Connecting links, supposed, 107.
Conscience, cases of, 201.
Conscientious Papuan, 197.
Cope, Professor, 71, 130.
Coracoid, of birds and reptiles, 70.
Cornea, 77.
Cornelius à Lapide, 265.
Correlation, laws of, 173.
Corti, fibres of, 53, 279.
Coryanthes, 56.
Costa, M., 88.
Cranial segments, 172.
Creation, 245, 252.
Creator, 15, 252.
Creed, Apostles', 245.
Crocodile, 43.
Croll, Mr., 137.
Crustacea, 79, 160.
Cryptacanthus, 146.
Crystalline matter, 266.
Crystals of snow, 186.
Cuttle-fishes, 74, 75.
Cuvier, 109.
Cyprinoids, 146.
Cytheridea, 79.

D.

DANA, Professor, 149.
Darwin, Mr. Charles, 2, 10, 12, 14—
21, 23, 27, 34, 35, 43, 45, 47, 48, 55
—57, 59, 65, 88, 94, 98—100,
107, 118—126, 129, 138, 142, 145,
149, 150, 181, 188—190, 196, 208,
209, 214—216, 218, 223, 233, 234,
252, 254, 258, 259, 275, 276.
Datura tatula, 101.
Delhi, days at, 98.
Delpino, Signor, 212, 213, 215.
Democritus, 217, 275, 288.
Density of air for breathing, 99.
Dentition, carnivorous, 110.
Derivation, 238.
Derivative creation, 252, 282.
Design, 259.
Devotion, 193.
Dibranchiata, 74.
Difficulties of problem of specific
origin, 1.
Digits, supernumerary, 122, 181.
Digits, turtle's, 106.
Dimorphodon, 71.
Dinornis, 70.
Dinosauria, 71.
Diseased pelvis, 182.
Dissemination of seeds, 65.
Doris, 170.
Dotheboys' Hall, 272.
Dragon, the flying, 64, 158.
Dragon-fly, 77.
Droughts, 25.
Duck-billed platypus, 175.
Dugong, 41, 175.
Duke of Argyll, 14, 276.
Dyspepsia, 201.

E.

EAR, 74.
Ear, formation of, 51.
Early specialization, 111.
Echinodermata, 44.
Echinoidea, 44.
Echinops, 148.
Echinorhinus, 172.
Echinus, 43.
Economy, Fuegian political, 192.
Eczema, 183.
Edentata, 174.
Egyptian monuments, 138.
Elasmobranchs, 140.
Elbow and knee affections, 183.

Empedocles, 288.
Eocene ungulata, 110.
Eolis, 170.
Equus, 97.
Ericulus, 148.
Ethics, 188.
Eudes Deslongchamps, 99.
Eurypterida, 141, 171.
Eutropius, 147.
Everett, Rev. R., 98.
Evolution requires geometrical in-
crease of time, 139.
Eye, 76.
Eye, formation of, 51.
Eye of trilobites, 135.

F.

FABRE, M., 46.
Feather-legged breeds, 181.
Feejeans, 199.
Fertilization of orchids, 55.
"Fiat justitia, ruat cœlum," 195.
Fibres of Corti, 53, 279.
Final misery, 194.
Finger of Potto, 105.
Fish, flying, 64.
Fishes, fresh-water, 145.
Fishes, thoracic and jugular, 39,
140.
Fixity of position of limbs, 39.
Flat-fishes, 37, 166.
Flexibility of bodily organization.
degrees of, 119.
Flexibility of mind, 267.
Flies, horned, 93.
Flight of spiders, 65.
Flounder, 37.
Flower, Professor, 163, 232, 283.
Fly, orchid, 55.
Flying-dragon, 64, 158.
Flying fish, 64.
Fœtal teeth of whales, 7.
Food, effects on pigs, 99.
Footsteps of Connecticut, 131.
Foraminifera, 186.
Formally moral acts, 195.
Formation of eye and ear, 51.
Forms, substantial, 186, 272.
Four-gilled Cephalopods, 76.
Fowls, white silk, 122.
French theatrical audience, 198.
Fresh-water fishes, 145.
Frogs, Chilian and European, 149.
Fuego, Terra del, 192.

G.

GALAGO, 158.
Galaxias, 147.
Galeus vulgaris, 172.
Galton, Mr. F., 97, 113, 228.
Gascoyen, Mr., 182.
Gavials, 43.
Gegenbaur, Prof., 176—178.
Gemmules, 208.
Generative system, its sensitiveness, 235.
Genesis of morals, 201.
Geographical distribution, 144.
Geographical distribution explained by Natural Selection, 6.
Geometrical increments of time, 139.
Geotria, 147.
Giraffe, neck of, 24.
Gizzard-like stomach, 83.
Glacial epoch, 150.
Glyptodon, 110.
Godron, Dr., 101.
Goose, its inflexibility, 119.
Göppert, Mr., 101.
Gould, Mr., 88.
Grasshopper, Great Shielded, 89.
Gray, Dr. Asa, 253, 255, 261.
Great Ant-eater, 83.
Great Salamander, 172.
Great Shielded Grasshopper, 89.
Greyhounds in Mexico, 99.
Greyhounds, time for evolution of, 138.
Guinea-fowl, 120.
Guinea-pig, 126.
Günther, Dr., 145, 146, 172.

H.

HAIRLESS DOGS, 174, 175.
Hamilton, Sir Wm., 267.
Harmony, musical, 54, 279.
Heart in birds and reptiles, 158.
Hegel, 217.
Heliconidæ, 29.
Hell, 194.
Heptanchus, 172.
Herbert Spencer, Mr., 20, 28, 67, 72, 163—166, 168, 170—172, 184, 187, 202, 203, 205, 218, 228, 245, 246, 248, 251.
Hessian flies, 170.
Heterobranchus, 146.
Hewitt, Mr., 124, 181.
Hexanchus, 172.

Hipparion, 97, 134.
Homogeny, 158.
Homology, bilateral or lateral, 156, 164.
Homology, meaning of term, 7, 156.
Homology, serial, 159.
Homology, vertical, 165.
Homoplasy, 159.
Honey-suckers, 90.
Hood of cobra, 50.
Hook-billed ducks, 100.
Hooker, Dr., 150.
Horned flies, 93.
Horny plates, 40, 42.
Horny stomach, 83.
Human larynx, 54, 278.
Humphry, Professor, 163.
Hutton, Mr. R. Holt, 202, 203.
Huxley, Professor, 67—69, 71, 72, 95, 103, 109, 130, 131, 137, 141, 163, 172, 173, 231, 247, 273.
Hybrids, mortality of, 124.
Hydrocyonina, 146.
Hyperphysical action, 253.
Hyrax, 179.

I.

ICHTHYOPSIDA, 109.
Ichthyosaurus, 78, 106, 132, 177.
Ichthyosis, 183.
Iguanodon, 71.
Illegitimate symbolic conceptions, 251.
Illustration by clock-thinking, 249.
Imaginal disks, 46, 170.
Implacental mammals, 67, 68.
Independent origins, 152.
Indian butterfly, 30.
Indian region's butterflies, 83.
Indians and cholera, 192.
Individual, meaning of word, 2.
Infirm, care of, 192.
Influence, local, 83.
Insect, walking-leaf, 35.
Insects, walking-stick and bamboo, 33.
Insectivora, 68.
Insectivorous mammals, 148.
Insectivorous teeth, 68.
Instinct of bull-dog, 260.
Intermediate forms, 128.
Intuitions, primary, 251.
Irregularities in blood-vessels, 182.
Isaria felina, 115.

J.

JAPANNED PEACOCK, 100.
Jews, 212.
Joints of backbone, 157, 162.
Jugular fishes, 39, 141.
Julia Pastrana, 174.

K.

KALLIMA INACHIS, 31.
Kallima paralekta, 31.
Kangaroo. 42, 67.
Kowalewsky, 81.
Knee and elbow affections, 183.
Kölliker, Professor, 104.

L.

LABYRINTHICI, 146.
Labyrinthodon, 104, 134.
Lamarck, 3.
Lankester, Mr. Ray, 152, 158.
Larynx of kangaroo, 42.
Larynx of man, 54, 278.
Lateral homology, 164.
Laws of correlation, 173.
Leaf butterfly, 31.
Legitimate symbolic conceptions, 251.
Lens, 76.
Lepidosteus, 172.
Lepra, 183.
Lewes, Mr. G. H., 94, 212, 214, 216.
Lewis, St., 206.
Lewis XV., 206.
Lewis XVI., 206.
Limb genesis, 176.
Limb muscles, 180.
Limbs, fixity of position of, 39.
Limbs of lobster, 161.
Links, supposed connecting, 107.
Lions, breeding. 234.
Lions, diseased pelvis, 182.
Llama, 109.
Local influences, 83.
Lobster, 160.
Long-tailed bird of Paradise, 91.
Lubbock, Sir John, 193, 204.
Lyell Sir, Charles, on dogs, 99, 106.

M.

MACHAIRODUS, 110.
Macrauchenia, 109.
Macropodidæ, 69.

Macroscelides, 68.
Madagascar, 148, 152.
Magnificent Bird of Paradise, 93.
Maize, American, 100.
Mammals, 67.
Mammary gland of kangaroo. 42.
Mammary gland, origin of, 47.
Man, origin of, 277.
Man reveals God, 267.
Man, voice of, 54.
Manatee, 41, 175.
Manchamp breed of sheep, 100.
Manis, 175.
Man's larynx, 54.
Many simultaneous modifications, 57.
Marcus Aurelius, 206.
Martineau, Mr. James, 200, 245.
Mastacembelus, 145.
Materially moral acts, 195.
Matter, crystalline and colloidal, 266.
Meaning of word "individual," 2.
Meaning of word "species," 2.
Mechanical theory of spine, 164.
Mediterranean oyster, 88, 98.
Meehan, Mr., 88.
Mexico, dogs in, 99.
Mill, John Stuart, 15, 189, 193, 194.
Mimicry, 8, 29.
Miracle, 287.
Molars, 111.
Mole, 176.
Molière, 230.
Mombas, cats at, 98.
Monkeys, American, 226.
Monster proboscis, 123.
Moral acts, 195.
Mordacia, 147.
Murphy, Mr. J. J., 52, 53, 76, 103, 114,
    115, 137, 185, 221, 276, 281.
Murray, Mr. Andrew, 83.
Mus delicatulus, 82.
Muscles of limbs, 180.
Mussel, 79.
Myrmecophaga, 83.

N.

NASALIS, SEMNOPITHECUS, 139.
Nathusius, 99.
Natural Selection, shortly stated, 5.
Naudin, M. C., 101.
Nautilus, 76.
Nebular evolution, 273.
Neck of giraffe, 24.
Newman, the Rev. Dr., 260, 268, 270.
    286.

New Zealand crustacea, 149.
New Zealand fishes, 147.
Niata cattle, 100.
Nile fishes, 146.
Normandy pig, 99.
North American fish, 147.
Nycticebus, 179.

O.

OBJECT of book, 5.
Objections from astronomy, 136.
Octopods, 77.
Offensive remarks of Prof. Vogt, 13.
Old, care of the, 192.
Old Fuegian women, 192.
Omygena exigua, 115.
Ophiocephalus, 146.
Optic lobes of pterodactyles, 71.
Orchids, 92.
Orchids, Bee, &c., 55.
Organ of hearing, 74.
Organ of sight, 76.
Organic polarities 185.
Origin of man, 277.
Orioles, 90.
Ornithoptera, 84.
Ornithorhynchus, 175.
Orthoceratidæ, 170.
Orycteropus 174.
Ostracods, 79.
Ostrich, 70.
Otoliths, 74.
Outlines of butterflies' wings, 86.
Owen, Professor, 74, 102, 123, 217, 238, 274.
Oyster of Mediterranean, 88, 98.
Oysters, 79.

P.

PAGET. Mr. J., 182.
Palæotherium, 109.
Pallas, 125.
Pangenesis, 19, 208.
Pangolin, 175.
Papilio Hospiton, 85.
Papilio Machaon, 85.
Papilio Ulysses, 84.
Papilionidæ, 83.
Papuan morals, 197, 198.
Parthenogenesis, 217.
Passiflora gracilis, 107.
Pastrana, Julia, 174.

Pathological polarities, 184.
Pavo nigripennis, 100.
Peacock, black shouldered, 100.
Peacock inflexibility of, 119.
Pedicellariæ, 44.
Pelvis, diseased, 182.
Pendulous appendages of turkey, 100.
Perameles, 68.
Periophthalmus, 146.
Perissodactyle ungulates, 109.
Permian, jugular fish, 141.
Perodicticus, 105, 179.
Phalangers, 67.
Phasmidæ, 89.
Phyllopods, 79.
Physical actions, 253.
" Physiological units," 168, 218.
Pigeons' "boots," 181.
Placental mammals, 67.
Placental reproduction, 81.
Plants. tendrils of, 107.
Plates of baleen, 40.
Platypus, 175.
Pleiades, 193.
Plesiosaurus, 106, 133, 178.
Pleurodont dentition, 148.
Pleuronectidæ, 37, 166.
Plotosus, 147.
Poisoning apparatus, 66.
Poisonous serpents, 50.
Polarities, organic, 184, 185.
Political economy, Fuegian, 192.
Polyzoa, 80, 81.
Pompadour, Madame de, 206.
Poppy, variety of, 101.
Porcupine, 175.
Porto Santo rabbit, 100, 122.
Potto, 105, 179.
Pouched beasts, 67.
Powell, the Rev. Baden, 259, 261, 285.
Premolars, 111.
Prepotency, 124.
Primary intuitions, 251.
Primitive man, 204.
Problem of origin of kinds, 1.
Proboscis monkey, 139.
Proboscis of ungulates, 123.
Processes, bird's-head, 80.
Psettus, 146.
Psoriasis, 183.
Pterodactyles, compared with birds, 70.
Pterodactyles, wing of, 64.
Puccinia, 115.
Purpose, 259.

Q.

QUASI-VERTEBRAL theory of skull,172.

R.

RABBIT of Porto Santo, 100, 122.
Radial ossicle, 176.
Rarefied air, effect on dogs, 99.
Rattlesnake, 49, 50.
Red bird of Paradise, 92.
Relations, analogical, 157.
Relations, homological, 156.
Reptiles compared with birds, 70.
Retina, 76.
Retrieving, virtue a kind of, 189, 205.
Reversion, cases of, 122.
Rhea, 70.
Ribs of Cetacea and Sirenia, 41.
Ribs of flying-dragon, 64, 158.
Richardson's figures of pigs, 99.
Roger Bacon, 266.
Rudimentary structures, 7, 102.

S.

SABRE-TOOTHED tiger, 110.
St. Augustin, 17, 263—265.
St. Basil, 17.
St. Hilaire, M., 179.
St. Thomas Aquinas, 17, 263, 265.
Salamander, great, 172.
Salter, Mr., 124.
Salvia officinalis, 213.
Salvia verticillata, 213.
Scapula of birds and reptiles, 70.
Schreber, 13.
Sclerotic, 76.
Scorpion, sting of, 66.
Seals, 83.
Sea squirts, 81.
Seeds, dissemination of, 65.
Seeley, Mr., on pterodactyles, 71.
Segmentation of skull, 172.
Segmentation of spine, 171.
Segments, similar, 160.
Self-existence, 252.
Semnopithecus, 139.
Sense, organ of, 51, 69, 74, 76.
Sensitiveness of generative system, 235.
Sepia, 77.
Serpents, poisonous, 50.
Sexual characters of apes, 49.
Sexual selection, 48.

Sharks, 83.
Shell-fish, beauty of; 54.
Shells of oysters, 88, 98.
Shielded grasshopper, 89.
Silurian strata, 140, 142.
Simultaneous modifications, 57.
Sirenia, 42.
Sir John Lubbock, 198, 204.
Sir William Thomson, 136.
Sitaris, 46.
Six-shafted bird of Paradise, 90.
Skull bones, 153.
Skull segments, 172.
Sloth, windpipe of, 82.
Smithfield, wife-selling in, 193.
Snow, crystals of, 186.
Sole, 37.
Solenodon, 148.
Species, meaning of word, 2.
Spelerpes, 165.
Spencer, see Herbert Spencer.
Spider orchid, 55.
Spiders, flight of, 65.
Spine of Glyptodon, 110.
Spine, segmentation of, 172.
Squalidæ, 38.
Squilla, 160.
Sterility of hybrids, 125.
Stings, 66.
Straining action of baleen, 41.
Struthious birds, 70, 151.
Sturgeon, 171.
Suarez, 18, 263.
Substantial forms, 186, 272.
Sufferings of beasts, 260.
Supernatural action, 252.
Supernatural action not to be looked for in nature, 15.
Supernumerary digits, 122, 181.
Syllis, 169, 211.
Symbolic conceptions, 251.
Symmetrical diseases, 182.
Syphilitic deposits, 183.

T.

TADPOLE's beak, 83.
Tails of butterflies, 85.
Tapir, 123, 134.
Tarsal bones, 159, 198.
Teeth of Cetacea, 83.
Teeth of Insectivora, 68.
Teeth of kangaroo and Macroscelides, 69.
Teeth of seals, 83.
Teeth of sharks, 83.

Teleology and evolution compatible, 273.
Tendrils of climbing plants, 107.
Tenia echinococcus, 170.
Teratology, 173.
Tetragonopterina, 146.
Thomson, Sir William, 136.
Thoracic fishes, 39.
Thorax of crustaceans, 79.
Thylacine, 67.
Tierra del Fuego, 192.
Tiger, sabre-toothed, 110.
Time required for evolution, 128.
Tope, 172.
Trabeculæ cranii, 172.
Transitional forms, 128.
Transmutationism, 242.
Trevelyan, Sir J. Peacock, 100.
Trilobites, 135, 141, 171.
Tunicaries, 81.
Turbot, 37.
Turkey, effects of climate on, 100.
Turkish dog, 45.
Two-gilled cephalopods, 76.
Type, conformity to, 241.

U.

UMBILICAL vesicle, 82.
Ungulata, 25, 109.
Ungulata eocene, 110.
Units, physiological, 168, 218.
Unknowable, the, 245.
Upper Silurian strata, 140, 142.
Urotrichus, 68.

V.

VARIABILITY, different degrees of, 119.
Vermiform appendix, 83.
Vertebræ of skull, 172.
Vertebral column, 162, 171.
Vertebrate limbs, 38, 163.
Vertical homology, 165.
Vesicle, umbilical, 82.
" Vestiges of Creation," 3.
View here advocated, 5.

Vitreous humour, 76.
Vogt, Professor, 12, 273.
Voice of man, 54.
Voltaire, 230.

W.

WAGNER, J. A., 13.
Wagner, Nicholas, 170.
Walking leaf, 35.
Walking-stick insect, 33.
Wallace, Mr. Alfred, 2, 10, 26, 29, 30, 32, 35, 36, 54, 83, 84, 87, 89, 90, 103, 117, 191, 197, 226, 274, 281—283.
Weaver fishes, 39.
Weitbrecht, 179.
Whale, fœtal teeth of, 7.
Whale, mouth of, 40.
Whalebone, 40.
Whales, 78.
White silk fowls, 122.
Wife selling, 198.
Wild animals, their variability, 120.
Wilder, Professor Burt, 180, 184.
Windpipe, 82.
Wings of bats, birds, and pterodactyles, 64, 130.
Wings of birds, origin of, 106.
Wings of butterflies, outline of, 86.
Wings of flying-dragon, 64, 158.
Wings of humming-bird, 157.
Wings of humming-bird hawk moth, 157.
Wings of insects, 65.
Wombat, 83.
Women, old Fuegian, 192.
Worms undergoing fission, 169, 211
Wyman, Dr. Jeffries, 185.

Y.

YORK MINSTER, a Fuegian, 197.

Z.

ZEBRAS, 134.
Zoological Gardens, Superintendent of, 126.

DARWINISM:

BEING

AN EXAMINATION OF MR. ST. GEORGE MIVART'S 'GENESIS OF SPECIES,'

[Reprinted from the 'North American Review,' July, 1871, with Additions.]

By CHAUNCEY WRIGHT, Esq.

LONDON:
JOHN MURRAY, ALBEMARLE STREET.
1871.

The right of Translation is reserved.

DARWINISM.

It is now nearly twelve years since the discussion of that
" mystery of mysteries," the origin of species, was reopened by
the publication of the first edition of Mr. Darwin's most remarkable work. Again and again in the history of scientific
debate this question had been discussed, and, after exciting a
short-lived interest, had been condemned by cautious and conservative thinkers to the limbo of insoluble problems or to
the realm of religious mystery. They had, therefore, sufficient
grounds, à priori, for anticipating that a similar fate would
attend this new revival of the question, and that, in a few
years, no more would be heard of the matter; that the same
condemnation awaited this movement which had overwhelmed
the venturesome speculations of Lamarck and of the author
of the 'Vestiges of Creation.' This not unnatural anticipation
has been, however, most signally disappointed. Every year has
increased the interest felt in the question, and at the present
moment a long list of recent publications testifies to the firm
hold which the subject has acquired in this short period on the
speculative interests of all inquisitive minds. But what can
we say has really been accomplished by this debate; and what
reasons have we for believing that the judgment of conservative
thinkers will not, in the main, be proved right after all, though
present indications are against them? One permanent consequence, at least, will remain, in the great additions to our
knowledge of natural history, and of general physiology, or
theoretical biology, which the discussion has produced; though
the greater part of this positive contribution to science is still
to be credited directly to Mr. Darwin's works, and even to his
original researches. But, besides this, an advantage has been
gained which cannot be too highly estimated. Orthodoxy has

been won over to the doctrine of evolution. In asserting this result, however, we are obliged to make what will appear to many persons important qualifications and explanations. We do not mean that the heads of leading religious bodies, even in the most enlightened communities, are yet willing to withdraw the dogma that the origin of species is a special religious mystery, or even to assent to the hypothesis of evolution as a legitimate question for scientific inquiry. We mean only, that many eminent students of science, who claim to be orthodox, and who are certainly actuated as much by a spirit of reverence as by scientific inquisitiveness, have found means of reconciling the general doctrine of evolution with the dogmas they regard as essential to religion. Even to those whose interest in the question is mainly scientific this result is a welcome one, as opening the way for a freer discussion of subordinate questions, less tramelled by the religious prejudices which have so often been serious obstacles to the progress of scientific researches.

But again, in congratulating ourselves on this result, we are obliged to limit it to the doctrine of evolution in its most general form, the theory common to Lamarck's zoölogical philosophy, to the views of the author of the ' Vestiges of Creation,' to the general conclusions of Mr. Darwin's and Mr. Wallace's theory of Natural Selection, to Mr. Spencer's general doctrine of evolution, and to a number of minor explanations of the processes by which races of animals and plants have been derived by descent from different ancestral forms. What is no longer regarded with suspicion as secretly hostile to religious beliefs by many truly religious thinkers is that which is denoted in common by the various names " transmutation," " development," "derivation," "evolution," and "descent with modification." These terms are synonymous in their primary and general signification, but refer secondarily to various hypotheses of the processes of derivation. But there is a choice among them on historical grounds, and with reference to associations, which are of some importance from a theological point of view. " Transmutation " and " development " are under ban. " Derivation " is, perhaps, the most innocent word; though " evolution " will probably prevail, since, spite of its etymological implication, it has lately become most acceptable, not only to the theological critics of the theory, but to its scientific advocates; although, from the neutral ground of experimental science, "descent with modification" is the most pertinent and least exceptionable name.

While the general doctrine of evolution has thus been successfully redeemed from theological condemnation, this is not yet true of the subordinate hypothesis of Natural Selection, to the partial success of which this change of opinion is, in great measure, due. It is, at first sight, a paradox that the views most peculiar to the eminent naturalist, whose work has been chiefly instrumental in effecting this change of opinion, should still be rejected or regarded with suspicion by those who have nevertheless been led by him to adopt the general hypothesis,— an hypothesis which his explanations have done so much to render credible. It would seem, at first sight, that Mr. Darwin has won a victory, not for himself, but for Lamarck. Transmutation, it would seem, has been accepted, but Natural Selection, its explanation, is still rejected by many converts to the general theory, both on religious and scientific grounds. But too much weight might easily be attributed to the deductive or explanatory part of the evidence, on which the doctrine of evolution has come to rest. In the half-century preceding the publication of the 'Origin of Species,' inductive evidence on the subject has accumulated, greatly outweighing all that was previously known; and the 'Origin of Species' is not less remarkable as a compend and discussion of this evidence than for the ingenuity of its explanations. It is not, therefore, to what is now known as "Darwinism" that the prevalence of the doctrine of evolution is to be attributed, or indirectly assigned. Still, most of this effect is due to Mr. Darwin's work, and something undoubtedly to the indirect influence of reasonings that are regarded with distrust by those who accept their conclusions; for opinions are contagious, even where their reasons are resisted.

The most effective general criticism of the theory of Natural Selection which has yet appeared, or one which, at least, is likely to exert the greatest influence in overcoming the remaining prejudice against the general doctrine of evolution, is the work of Mr. St. George Mivart 'On the Genesis of Species.' Though, as we shall show in the course of this article, the work

falls far short of what we might have expected from an author
of Mr. Mivart's attainments as a naturalist, yet his position
before the religious world, and his unquestionable familiarity
with the theological bearings of his subject, will undoubtedly
gain for him and for the doctrine of evolution a hearing and a
credit, which the mere student of science might be denied. His
work is mainly a critique of "Darwinism;" that is, of the
theories peculiar to Mr. Darwin and the "Darwinians," as distinguished from the believers in the general doctrine of evolution which our author accepts. He also puts forward an
hypothesis in opposition to Mr. Darwin's doctrine of the predominant influence of Natural Selection in the generation of
organic species, and their relation to the conditions of their
existence. On this hypothesis, called "Specific Genesis," an
organism, though at any one time a fixed and determinate
species, approximately adapted to surrounding conditions of
existence, is potentially, and by innate potential combinations
of organs and faculties, adapted to many other conditions of
existence. It passes, according to the hypothesis, from one
form to another of specific "manifestation," abruptly and dis-
continuously in conformity to the emergencies of its outward
life; but in any condition to which it is tolerably adapted it
retains a stable form, subject to variation only within determinate limits, like oscillations in a stable equilibrium. For
this conception our author is indebted to Mr. Galton, who, in
his work on 'Hereditary Genius,' "compares the development
of species with a many-faceted spheroid tumbling over from
one facet or stable equilibrium to another. The existence of
internal conditions in animals," Mr. Mivart adds (p. 111),
" corresponding with such facets is denied by pure Darwinians,
but it is contended in this work that something may also be
said for their existence." There are many facts of variation,
numerous cases of abrupt changes in individuals both of natural
and domesticated species, which, of course, no Darwinian or
physiologist denies, and of which Natural Selection professes to
offer no direct explanation. The causes of these phenomena,
and their relations to external conditions of existence, are
matters quite independent of the principle of Natural Selection,
except so far as they may directly affect the animal's or plant's

well-being, with the origin of which this principle is alone concerned. General physiology has classified some of these sudden variations under such names as "reversion" and "atavism," or returns more or less complete to ancestral forms. Others have been connected together under the law of "correlated or concomitant variations," changes that, when they take place, though not known to be physically dependent on each other, yet usually or often occur together. Some cases of this law have been referred to the higher, more fundamental laws of homological variations, or variations occurring together on account of the relationships of homology, or due to similarities and physical relations between parts of organisms, in tissues, organic connections, and modes of growth. Other variations are explained by the laws and causes that determine monstrous growths. Others again are quite inexplicable as yet, or cannot yet be referred to any general law or any known antecedents. These comprise, indeed, the most common cases. The almost universal prevalence of well-marked phenomena of variation in species, the absolutely universal fact that no two individual organisms are exactly alike, and that the description of a species is necessarily abstract and in many respects by means of averages,—these facts have received no particular explanations, and might indeed be taken as ultimate facts or highest laws in themselves, were it not that in biological speculations such an assumption would be likely to be misunderstood, as denying the existence of any real determining causes and more ultimate laws, as well as denying any known antecedents or regularities in such phenomena. No physical naturalist would for a moment be liable to such a misunderstanding, but would, on the contrary, be more likely to be off his guard against the possibility of it in minds otherwise trained and habituated to a different kind of studies. Mr. Darwin has undoubtedly erred in this respect. He has not in his works repeated with sufficient frequency his faith in the universality of the law of causation, in the phenomena of general physiology or theoretical biology, as well as in all the rest of physical nature. He has not said often enough, it would appear, that in referring any effect to " accident," he only means that its causes are like particular phases of the weather, or like innumerable phenomena in the concrete course of nature generally, which are quite beyond the power of finite minds to anticipate or to account for in detail, though none the less really determinate or due to regular causes. That he has committed this error appears from the fact that his critic, Mr. Mivart, has made the mistake, which nullifies nearly the whole of his criticism, of supposing that "the theory of Natural Selection may (though it need not) be taken in such a way as to lead men to regard the present organic world as formed, so to speak, *accidentally,* beautiful and wonderful as is confessedly the haphazard result" (p. 33). Mr. Mivart, like many another writer, seems to forget the age of the world in which he lives and for which he writes,—the age of " experimental philosophy," the very stand-point of which, its fundamental assumption, is the universality of physical causation. This is so familiar to minds bred in physical studies, that they rarely imagine that they may be mistaken for disciples of Democritus, or for believers in "the fortuitous concourse of atoms," in the sense, at least, which theology has attached to this phrase. If they assent to the truth that may have been meant by the phrase, they would not for a moment suppose that the atoms move fortuitously, but only that their conjunctions, constituting the actual concrete orders of events, could not be anticipated except by a knowledge of the natures and regular histories of each and all of them,—such knowledge as belongs only to omniscience. The very hope of experimental philosophy, its expectation of constructing the sciences into a true philosophy of nature, is based on the induction, or, if you please, the *à priori* presumption, that physical causation is universal; that the constitution of nature is written in its actual manifestations, and needs only to be deciphered by experimental and inductive research; that it is not a latent invisible writing, to be brought out by the magic of mental anticipation or metaphysical meditation. Or, as Bacon said, it is not by the " anticipations of the mind," but by the "interpretation of nature," that natural philosophy is to be constituted; and this is to presume that the order of nature is decipherable, or that causation is everywhere either manifest or hidden, but never absent.

Mr. Mivart does not wholly reject the process of Natural Selection, or disallow it as a real cause in nature, but he reduces it to "a subordinate rôle" in his view of the derivation of species. It serves to perfect the imperfect adaptations and to meet within certain limits unfavourable changes in the conditions of existence. The "accidents" which Natural Selection acts upon are allowed to serve in a subordinate capacity and in subjection to a foreordained, particular, divine order, or to act like other agencies dependent on an evil principle, which are compelled to turn evil into good. Indeed, the only difference on purely scientific grounds, and irrespective of theological considerations, between Mr. Mivart's views and Mr. Darwin's is in regard to the *extent* to which the process of Natural Selection has been effective in the modifications of species. Mr. Darwin himself, from the very nature of the process, has never supposed for it, as a cause, any other than a co-ordinate place among other causes of change, though he attributes to it a superintendent, directive, and controlling agency among them. The student of the theory would gather quite a different impression of the theory from Mr. Mivart's account of it, which attributes to "Darwinians" the absurd conception of this cause as acting " alone " to produce the changes and stabilities of species; whereas, from the very nature of the process, other causes of change, whether of a known or as yet unknown nature, are presupposed by it. Even Mr. Galton's and our author's hypothetical "facets," or internal conditions of abrupt changes and successions of stable equilibriums, might be among these causes, if there were any good inductive grounds for supposing their existence. Reversional and correlated variations are, indeed, due to such internal conditions and to laws of inheritance, which have been ascertained inductively as at least laws of phenomena, but of which the causes, or the antecedent conditions in the organism, are unknown. Mr. Darwin continually refers to variations as arising from unknown causes, but these are always such, so far as observation can determine their relations to the organism's conditions of existence, that they are far from accounting for, or bearing any relations to, the adaptive characters of the organism. It is solely upon and with reference to such adaptive characters that the process of Natural Selection has any agency, or could be supposed to be effective. If Mr. Mivart had cited anywhere in his book, as he has not, even a single instance of sudden variation in a whole race, either in a state of nature or under domestication, which is not referable by known physiological laws to the past history of the race on the theory of evolution, and had further shown that such a variation was an adaptive one, he might have weakened the arguments for the agency and extent of the process of Natural Selection. As it is, he has left them quite intact.

The only direct proofs which he adduces for his theory that adaptive as well as other combinations proceed from innate pre-determinations wholly within the organism, are drawn from, or rather assumed in, a supposed analogy of the specific forms in organisms to those of crystals. As under different circumstances or in different media the same chemical substances or constituent substances assume different and distinct crystalline forms, so, he supposes, organisms are distinct manifestations of typical forms, one of which after another will appear under various external conditions. He quotes from Mr. J. J. Murphy, 'Habit and Intelligence,' that "it needs no proof that in the case of spheres and crystals, the forms and structures are the effect and not the cause of the formative principle. Attraction, whether gravitative or capillary, produces the spherical form; the spherical form does not produce attraction. And crystalline polarities produce crystalline structure and form; crystalline structure and form do not produce polarities." And, by analogy, Mr. Murphy and our author infer that innate vital forces always produce specific vital forms, and that the vital forms themselves, or " accidental " variations of them, cannot modify the types of action in vital force. Now, although Mr. Murphy's propositions may need no proof, they will bear correction; and clear as they appear to be, a better interpretation of the physical facts is needed for the purposes of tracing out analogy and avoiding paralogism. Strange as it may seem, Mr. Murphy's clear antitheses are not even partially true. No abstraction ever produced any other abstraction, much less a concrete thing. The abstract laws of attraction never produced any body, spherical or polyhedral. It was actual forces acting in definite ways that made the sphere or crystal; and the sizes, particular shapes, and positions of these bodies determined in part the action of these actual forces. It is the resultants of many actual attractions, dependent in turn on the actual products, that determine the spherical or crystalline forms. Moreover, in the case of crystals, neither these forces nor the abstract law of their action in producing definite angles reside in the finished bodies, but in the properties of the surrounding media, portions of whose constituents are changed into crystals, according to these properties and to other conditioning circumstances. So far as these bodies have any innate principle in them concerned in their own production, it is manifested in determining, not their general agreements, but their particular differences in sizes, shapes, and positions. The particular position of a crystal that grows from some fixed base or nucleus, and the particular directions of its faces, may, perhaps, be said to be *innate*; that is, they were determined at the beginning of the particular crystal's growth. Finding, therefore, what Mr. Murphy and Mr. Mivart suppose to be innate to be really in the outward conditions of the crystal's growth, and what they would suppose to be superinduced to be all that is innate in it, we have really found the contrast in place of an analogy between a crystal and an organism. For, in organisms, no doubt, and as we may be readily convinced without resort to analogy, there is a great deal that is really innate, or dependent on actions in the organism, which diversities of external conditions modify very little, or affect at least in a very indeterminate manner, so far as observation has yet ascertained. External conditions are, nevertheless, essential factors in development, as well as in mere increase or growth. No animal or plant is developed, nor do its developments acquire any growth without very special external conditions. These are quite as essential to the production of an organism as a crystalline nucleus and fluid material are to the growth and particular form of a crystal; and as the general resemblances of the crystals of any species, the agreements in their angles, are results of the physical properties of their food and other surrounding conditions of their growth, so the general resemblances of animals or plants of any species, their agreements in specific characters, are doubtless due, in the main, to the properties of what is innate in them, yet not to any abstraction. This is sufficiently conspicuous not to " need any proof," and is denied by no Darwinian. The analogy is so close indeed between the internal determinations of growth in an organism and the external ones of crystals, that Mr. Darwin was led by it to invent his "provisional hypothesis of Pangenesis," or theory of gemmular reproduction. The gemmules in this theory being the perfect analogues of the hypothetical atoms of the chemical substances that are supposed to arrange themselves in crystalline forms, the theory rather gives probability to the chemical theory of atoms than borrows any from it. But we shall recur to this theory of Pangenesis further on.

General physiology, or physical and theoretical biology, are sciences in which, through the study of the laws of inheritance, and the direct and indirect effect of external conditions, we must arrive, if in any way, at a more and more definite knowledge of the causes of specific manifestations; and this is what Mr. Darwin's labours have undertaken to do, and have partially accomplished. Every step he has taken has been in strict conformity to the principles of method which the examples of inductive and experimental science have established. A stricter observance of these by Mr. Murphy and our author might have saved them from the mistake we have noticed, and from many others,—the "realism" of ascribing efficacy to an abstraction making attraction and polarity produce structures and forms, independently of the products and of the concrete matters and forces in them. A similar "realism" vitiates nearly all speculations in theoretical biology, which are not designedly, or even instinctively, as in Mr. Darwin's work, made to conform to the rigorous rules of experimental philosophy. These require us to assume no causes that are not true or phenominally known, and known in some other way than in the effect to be explained; and to prove the sufficiency of those we do assume in some other way than by putting an abstract name or description of an effect for its cause, like using the words "attraction" and "polarity" to account for things the matters of which have *come together* in a *definite form.* It may seem strange to many readers to be told that Mr. Darwin, the most consummate speculative genius of our times, is no more a maker of hypothesis than Newton was, who, unable to discover the cause of the properties of gravitation, wrote the often-quoted but much misunderstood words, "*Hypotheses non fingo.*" "For," he adds, "whatever is not deduced from the phenomena is to be called an hypothesis; and hypotheses, whether metaphysical or physical, whether of occult qualities or mechanical, have no place in experimental philosophy. In this philosophy particular propositions are inferred from the phenomena, and afterwards rendered general by induction. Thus it was that the impenetrability, the mobility, and the impulsive force of bodies, and the laws of motion and gravitation were discovered. And to us it is enough that gravity does really exist and act according to the laws which we have explained, and abundantly serves to account for all the motions of the celestial bodies and of our sea." Thus, also, it is that the variability of organisms and the known laws of variation and inheritance, and of the influences of external conditions, and the law of Natural Selection, have been discovered. And though it is not enough that variability and selection do really exist and act according to laws which Mr. Darwin has explained (since the limits of their action and efficiency are still to be ascertained), yet it is enough for the present that Darwinians do not rest, like their opponents, contented with framing what Newton would have called, if he had lived after Kant, "*transcendental hypotheses,*" which have no place in experimental philosophy. It may be said that Mr. Darwin has invented the hypothesis of Pangenesis, against the rules of this philosophy; and so also did Newton invent the corpuscular theory of light, with a similar purpose and utility.

In determining the limits of the action of Natural Selection, and its sufficiency within these limits, the same demonstrative adequacy should not, for obvious reasons, be demanded as conditions of assenting to its highly probable truth, that Newton proved for his speculation. For the facts for this investigation are hopelessly wanting. Astronomy presents the anomaly, among the physical sciences, of being the only science that deals in the concrete with a few naturally isolated causes, which are separated from all other lines of causation in a way that in other physical sciences can only be imitated in the carefully guarded experiments of physical and chemical laboratories. The study of animals and plants under domestication is, indeed, a similar mode of isolating with a view to ascertaining the physical laws of life by inductive investigations. But the theory of Natural Selection, in its actual application to the phenomena of life and the origin of species, should not be compared to the theory of gravitation in astronomy, nor to the principles of physical science as they appear in the natures that are shut in by the experimental resources of the laboratory, but rather to these principles as they are actually working, and have been working, in the concrete courses of outward nature, in meteorology and physical geology. Still better, perhaps, at least for the purposes of illustration, we may compare the principle of Natural Selection to the fundamental laws of political economy, demonstrated and actually at work in the production of the values and the prices in the market of the wealth which human needs and efforts demand and supply. Who can tell from these principles what the market will be next week, or account for its prices of last week, even by the most ingenious use of hypotheses to supply the missing evidence. The empirical economist and statistician imagines that he can discover some other principles at work, some predetermined regularity in the market, some " innate " principles in it, to which the general laws of political economy are subordinated; and speculating on them, might risk his own wealth in trade, as the speculative " vitalist " might, if anything could be staked on a transcendental hypothesis. In the same way the empirical weather-philosopher thinks he can discern regularities in the weather, which the known principles of mechanical and chemical physics will not account for, and to which they are subordinate. This arises chiefly from his want of imagination, of a clear mental grasp of these principles, and of an adequate knowledge of the resources of legitimate hypothesis to supply the place of the unknown incidental causes through which these principles act. Such are also the sources of most of the difficulties which our author has found in the applications of the theory of Natural Selection.

His work is chiefly taken up with these difficulties. He does not so much insist on the probability of his own transcendental hypothesis, as endeavour to make way for it by discrediting the sufficiency of its rival ; as if this could serve his purpose ;

as if experimental philosophy itself, without aid from " Darwinism," would not reject his metaphysical, occult, transcendental hypothesis of a specially predetermined and absolute fixity of species,—an hypothesis which multiplies species in an organism to meet emergencies,—the emergencies of theory,—much as the epicycles of Ptolemy had to be multiplied in the heavens. Ptolemy himself had the sagacity to believe that his was only a mathematical theory, a mode of representation, not a theory of causation; and to prize it only as representative of the facts of observation, or as " saving the appearances." Mr. Mivart's theory, on the other hand, is put forward as a theory of causation, not to save appearances, but to justify the hasty conclusion that they are real; the appearances, namely, of complete temporary fixity, alternating with abrupt changes, in the forms of life which are exhibited by the scanty records of geology and in present apparently unchanging natural species.

Before proceeding to a special consideration of our author's difficulties on the theory of Natural Selection, we will quote from Mr. Darwin's latest work, ' The Descent of Man,' his latest views of the extent of the action of this principle and its relations to the general theory of evolution. He says (Chapter IV.) :—

" Thus a very large yet undefined extension may safely be given to the direct and indirect results of Natural Selection; but I now admit, after reading the essay by Nägeli on plants, and the remarks by various authors with respect to animals, more especially those recently made by Professor Broca, that in the earlier editions of my ' Origin of Species ' I probably attributed too much to the action of Natural Selection, or the survival of the fittest. I have altered the fifth edition of the ' Origin ' [the edition which Mr. Mivart reviews in his work], so as to confine my remarks to adaptive changes of structure. I had not formerly sufficiently considered the existence of many structures which appear to be, as far as we can judge, neither beneficial nor injurious; and this I believe to be one of the greatest oversights as yet detected in my work. I may be permitted to say, as some excuse, that I had two distinct objects in view: firstly, to show that species had not been separately created; and secondly, that Natural Selection had been the chief agent of change, though largely aided by the inherited effects of habit, and slightly by the direct action of the surrounding conditions. Nevertheless, I was not able to annul the influence of my former belief, then widely prevalent, that each species had been purposely created; and this led to my tacitly assuming that every detail of structure, excepting rudiments, was of some special, though unrecognized, service. Any one with this assumption in his mind would naturally extend the action of Natural Selection, either during past or present times, too far. Some of those who admit the principle of evolution, but reject Natural Selection, seem to forget when criticising my work, that I had the above two objects in view; hence, if I have erred in giving to Natural Selection great power, which I am far from admitting, or in having exaggerated its power, which is in itself probable, I have at least, as I hope, done good service in aiding to overthrow the dogma of separate creations."

In one other respect Mr. Darwin has modified his views of the action of Natural Selection, in consequence of a valuable criticism in the 'North British Review' of June, 1867; and our author regards this modification as very important, and says of it that "this admission seems almost to amount to a change of front in the face of the enemy." It is not, as we shall see, an important modification at all, and does not change in any essential particular the theory as propounded in the first edition of the 'Origin of Species,' but our author's opinion of it has helped us to discover what, without this confirmation, seemed almost incredible, how completely he has misapprehended, not merely the use of the theory in special applications, which is easily excusable, but also the nature of its general operation and of the causes employed by it; thus furnishing an additional illustration of what he says in his Introduction, that "few things are more remarkable than the way in which it [this theory] has been misunderstood." One other consideration has also been of aid to us. In his concluding chapter on "Theology and Evolution," in which he very ably shows, and on the most venerable authority, that there is no necessary conflict between the strictest orthodoxy and the theory of evolution, he remarks (and quotes Dr. Newman) on the narrowing effect of single lines of study. Not only inabilities may be produced by a one-sided pursuit, but " a positive distaste may grow up, which, in the intellectual order, may amount to a spontaneous and unreasoning disbelief in that which appears to be in opposition to the more familiar concept, and this at all times." This is, of course, meant to apply to those who, from want of knowledge, also lack ability and interest and even acquire a distaste for theological studies. But it also has other and equally important applications. Mr. Mivart, it would at first sight seem, being distinguished as a naturalist and also versed in theology, is not trammelled by any such narrowness as to disable him from giving just weight to both sides of the question he discusses. But what are the two sides? Are they the view of the theologian and the naturalist? Not at all. The debate is between the theologian and descriptive naturalist on one side, or the theologian and the student of natural history in its narrowest sense, that is, systematic biology; and on the other side the physical naturalist, physiologist, or theoretical biologist. Natural history and biology, or the general science of life, are very comprehensive terms, and comprise in their scope widely different lines of pursuit and a wide range of abilities. In fact, the sciences of biology contain contrasts in the objects, abilities, and interests of scientific pursuit almost as wide as that presented by the physical sciences generally, and the sciences of direct observation, description, and classification. The same contrast holds, indeed, even in a science so limited in its material objects as astronomy. The genius of the practical astronomer and observer is very different from that of the physical astronomer and mathematician; though success in this science generally requires now-a-days that some degree of both should be combined. So the genius of the physiologist is different from that of the naturalist proper, though in the study of comparative anatomy the observer has to exercise some of the skill in analysis and in the use of hypotheses which are the genius of the physical sciences in the search for unknown causes. We may, perhaps, comprise all the forms of intellectual genius (excluding æsthetics) under three chief classes, namely, first, the genius that pursues successfully the researches for unknown causes by the skilful use of hypothesis and experiment; secondly, that which, avoiding the use of hypotheses or preconceptions altogether and the delusive influence of names, brings together in clear connections and contrasts in classification the objects of nature in their broadest and realest relations of resemblance ; and thirdly, that genius which seeks with success for reasons and authorities in support of cherished convictions.

That our author may have the last two forms of genius, even in a notable degree, we readily admit ; but that he has not the first to the degree needed for an inquiry, which is essentially a branch of physical science, we propose to show. We have already pointed out how his theological education, his schooling against Democritus, has misled him in regard to the meaning of " accidents " or accidental causes in physical science; as if to the physical philosopher these could possibly be an absolute and distinct class, not included under the law of causation, " that every event must have a cause or determinate antecedents," whether we can trace them out or not. The accidental causes of science are only " accidents " relatively to the intelligence of a man. Eclipses have the least of this character to the astronomer of all the phenomena of nature ; yet to the savage they are the most terrible of monstrous accidents. The accidents of monstrous variation, or even of the small and limited variations normal in any race or species, are only accidents relatively to the intelligence of the naturalist, or to his knowledge of general physiology. An accident is what cannot be anticipated from what we know, or by any intelligence, perhaps, which is less than omniscient.

But this is not the most serious misconception of the accidental causes of science which our author has fallen into. He utterly mistakes the particular class of accidents concerned in the process of Natural Selection. To make this clear, we will enumerate the classes of causes which are involved in this process. In the first place, there are the external conditions of an animal's or plant's life, comprising chiefly its relations to other organic beings, but partly its relations to inorganic nature, and determining its needs and some of the means of satisfying them. These conditions are consequences of the external courses of events or of the partial histories of organic and inorganic nature. In the second place, there are the general principles of the fitness of means to ends, or of supplies to needs. These comprise the best ascertained and most fundamental of all the principles of science, such as the laws of mechanical, optical, and acoustical science, by which we know how a leg, arm, or wing, a bony frame, a muscular or a vascular system, an eye or an ear, can be of *use*. In the third place there are the causes introduced by Mr. Darwin to the attention of physiologists, as normal facts of organic nature, the little known phenomena of variation, and their relations to the laws of inheritance. There are several classes of these. The most important in the theory of Natural Selection are the diversities *always existing* in any race of animals or plants, called "individual differences," which *always* determine a better fitness of some individuals to the general conditions of the existence of a race than other less fortunate individuals have. The more than specific agreements in characters, which the best fitted individuals of a race must thus exhibit, ought, if possible, according to Cuvier's principles of Zoölogy, to be included in the description of a species (as a norm or type which only the *best* exhibit), instead of the rough averages to which the naturalist really resorts in defining species by marks or characters that are variable. But probably such averages in variable characters are really close approximations to the characters of the best general adaptation; for variation being, so far as known, irrespective of adaptation, is as likely to exist to the same extent on one side of the norm of utility as on the other, or by excess as generally as by defect. Though variation is irrespective of utility, its limits are not. Too great a departure from the norm of utility must put an end to life and its successions. Utility therefore determines, along with the laws of inheritance, not only the middle line or safest way of a race, but also the bounding limits of its path of life; and so long as the conditions and principles of utility embodied in a form of life remain unchanged, they will, together with the laws of inheritance, maintain a race unchanged in its average characters. "Specific stability," therefore, for which theological and descriptive naturalists have speculated a transcendental cause, is even more readily and directly accounted for by the causes which the theory of Natural Selection regards than is specific change. But just as obviously it follows from these causes that a change in the conditions and resources of utility, not only may but must change the normal characters of a species, or else the race must perish. Again, a slow and gradual change in the conditions of existence must, on these principles, slowly change the middle line or safest way of life (the descriptive or graphic line); but always, of course, this change must be within the existing limits of variation, or the range of "individual differences." A change in these limits would then

follow, or the range of "individual differences" would be extended, at least, so far as we know, in the direction of the
change. That it is widened or extended to a greater range by
rapid and important changes in conditions of existence, is a
matter of observation in many races of animals and plants that
have been long subject to domestication, or to the capricious
conditions imposed by human choice and care. This phenomenon is like what would happen if a roadway or path across
a field were to become muddy or otherwise obstructed. The
travelled way would swerve to one side, or be broadened, or
abandoned, according to the nature and degree of the obstruc-
tion, and to the resources of travel that remained. This class
of variations, that is "individual differences," constant and
normal in a race, but having different ranges in different races,
or in the same race under different circumstances, may be
regarded as in no proper sense accidentally related to the
advantages that come from them; or in no other sense than
a tendril, or a tentacle, or a hand searching in the dark, is
accidentally related to the object it succeeds in finding. And
yet we say properly that it was by "accident" that a certain
tendril was put forth so as to fulfil its function, and clasp the
particular object by which it supports the vine; or that it was
an accidental movement of the tentacle or hand that brought
the object it has secured within its grasp. The search was, and
continues to be, normal and general; it is the particular success
only that is accidental; and this only in the sense that lines
of causation, stretching backwards infinitely and unrelated
except in a first cause, or in the total order of nature, come
together, and by their concurrence produce it. Yet over even
this concurrence "law" still presides, to the effect that for
every such concurrence the same consequences follow.

But our author, with his mind filled with horror of "blind
chance," and of "the fortuitous concourse of atoms," has
entirely overlooked the class of accidental variations, on which,
even in the earlier editions of the 'Origin of Species,' the
theory of Natural Selection is based, and has fixed his attention exclusively on another class, namely abnormal or unusual
variations, which Mr. Darwin at first supposed might also be
of service in this process. The fault might, perhaps, be charged

against Mr. Darwin for not sufficiently distinguishing the two classes, as well as overlooking, until it was pointed out by his critic in the 'North British Review,' before referred to, the fact that the latter class could be of no service; if it were not that our author's work is a review of the last edition of the 'Origin of Species,' and of the treatise on 'Animals and Plants under Domestication,' in both of which Mr. Darwin has emphatically distinguished these classes, and admitted that it is upon the first class only that Natural Selection can normally depend; though the second class of unusual and monstrous variations may give rise, by highly improbable though possible accidents, to changes in the characters of whole races. Mr. Mivart characterizes this admission by the words we have quoted, that " it seems almost to amount to a change of front in the face of the enemy;" of which it might have been enough to say, that the strategy of science is not the same as that of rhetorical disputation, and aims at cornering facts, not antagonists. But Mr. Mivart profits by it as a scholastic triumph over heresy, which he insists upon celebrating, rather than as a correction of his own misconceptions of the theory. He continues throughout his book to speak of the variations on which Natural Selection depends as if they were all of rare occurrence, like abrupt and monstrous variations, instead of being always present in a race; and also as having the additional disadvantage of being " individually slight," " minute," " insensible," " infinitesimal," " fortuitous," and " indefinite." These epithets are variously combined in different passages, but his favourite compendious formula is " minute, fortuitous, and indefinite variations." When, however, he comes to consider the enormous time which such a process must have taken to produce the present forms of life, he brings to bear all his forces, and says (p. 154): " It is not easy to believe that less than two thousand million years would be required for the totality of animal development by no other means than minute, fortuitous, occasional, and intermitting variations in all conceivable directions." This exceeds very much—by some two hundred fold—the length of time Sir William Thomson allows for the continuance of life on the earth. It is difficult to see how, with such uncertain " fortuitous, occasional, and intermitting" elements, our author could have succeeded in making any calculations at all. On the probability of the correctness of Sir William Thomson's physical arguments, " the author of this book cannot presume to advance an opinion; but," he adds (p. 150), " the fact that they have not been refuted pleads strongly in their favour, when we consider how much they tell against the theory of Mr. Darwin." He can, it appears, judge of them on his own side.

For the descriptive epithets which our author applies to the variations on which he supposes Natural Selection to depend, he has the following authority. He says (p. 35) : " Now it is distinctly enunciated by Mr. Darwin that the spontaneous varia- tions upon which his theory depends are individually slight, minute, and insensible. He says ('Animals and Plants under Domestication,' Vol. II. p. 192): ' Slight individual differences, however, suffice for the work, and are probably the sole differ- ences which are effective in the production of new species.' " After what we have said as to the real nature of the differences from which nature selects, it might be, perhaps, unnecessary to explain what ought at least to have been known to a natu- ralist, that by " individual differences " is meant the differences between the individuals of a race of animals or plants; that the slightness of them is only relative to the differences between the characters of species, and that they may be very consider- able in themselves, or their effects, or even to the *eye* of the naturalist. How the expression "slight individual differences " could have got translated in our author's mind into "indivi- dually slight, minute, and insensible " ones, has no natural explanation. But this is not the only instance of such an un- fathomable translation in our author's treatment of the theory of Natural Selection. Two others occur on page 133. In the first he says : " Mr. Darwin abundantly demonstrates the varia- bility of dogs, horses, fowls, and pigeons, but he none the less shows the *very small* extent to which the goose, the peacock, and the guinea-fowl have varied. Mr. Darwin attempts to explain this fact as regards the goose by the animal being valued only for food and feathers, and from no pleasure having been felt in it on other accounts. He adds, however, at the end, the striking remark, which concedes the whole position,

' but the goose seems to have *a singularly inflexible organization.*' " The translation is begun in the author's italics, and completed a few pages further on (p. 141), where, recurring to this subject, he says: "We have seen that Mr. Darwin himself implicitly admits the principle of specific stability in asserting the singular inflexibility of the organization of the goose." This is what is called in scholastic logic, *Fallacia a dicto secundum quid ad dictum simpliciter.* The obvious meaning, both from the contexts and the evidence, of the expression " singularly inflexible," is that the goose has been much less changed by domestication than other domestic birds. But this *relative* inflexibility is understood by our author as an admission of an *absolute* one, in spite of the evidence that geese have varied from the wild type, and have individual differences, and even differences of breeds, which are sufficiently conspicuous, even to the eye of a goose. The next instance of our author's translations (p. 133) is still more remarkable. He continues: " This is not the only place in which such expressions are used. He [Mr. Darwin] elsewhere makes use of phrases which quite harmonize with the conception of a normal specific constancy, but varying greatly and suddenly at intervals. Thus he speaks of a *whole organism seeming to have become plastic and tending to depart from the parental type* ('Origin of Species,' 5th edit., 1869, p. 13)." The italics are Mr. Mivart's. The passage from which these words are quoted (though they are not put in quotation marks) is this: "It is well worth while carefully to study the several treatises on some of our old cultivated plants, as on the hyacinth, potato, even the dahlia, etc.; and it is really surprising to note the endless points in structure and constitution in which the varieties and sub-varieties differ slightly from each other. The whole organization seems to have become plastic, and tends to depart *in a slight degree* from that of the parental type." The words that we have italicized in this quotation are omitted by our author, though essential to the point on which he cites Mr. Darwin's authority, namely, as to the organism "varying greatly and suddenly at intervals." Logic has no adequate name for this fallacy; but there is another in our author's understanding of the passage which is very familiar,—the fallacy of ambiguous terms. Mr.

Darwin obviously uses the word " plastic " in its secondary signification as the name of that which is "capable of being moulded, modelled, or fashioned to the purpose, as clay." But our author quite as obviously understands it in its primary signification as the name of anything " having the power to give form." But this is a natural enough misunderstanding, since in scholastic philosophy the primary signification of "plastic" is the prevailing one.

Such being our author's misconceptions of the principle of Natural Selection, and such their source, it would be useless to follow him in his tests of it by hypothetical illustrations from the history of animals; but we are bound to make good our assertion that the author's difficulties have arisen, not only from his want of a clear mental grasp of principles, but also from an inadequate knowledge of the resources of legitimate hypothesis to supply the unknown incidental causes through which the principle has acted. These deficiencies of knowledge and imagination, though more excusable, are not less conspicuous in his criticisms than the defects we have noticed. He says (p. 59): " It may be objected, perhaps, that these difficulties are difficulties of *ignorance;* that we cannot explain them, because we do not know *enough* of the animals." It is not surprising that he adds : " But it is here contended that this is not the case; it is not that we merely fail to see how Natural Selection acted, but that there is a positive incompatibility between the cause assigned and the results." And no wonder that he remarks at the close of the chapter (Chapter II.) : " That minute, fortuitous, and indefinite variations could have brought about such special forms and modifications as have been enumerated in this chapter seems to contradict, not imagination, but reason."

In this chapter on 'Incipient Structures,' the fact is quite overlooked, which is so conspicuous in the principles of comparative anatomy, how few the fundamental structures are, which have been turned to such numerous uses; how meagre have been the resources of Natural Selection, so far as it has depended on the occurrence of structures which were of no previous use, or were not already partially useful in directions in which they have been modified by the selection and inheritance of "individual differences ; " or how important to Natural Selection have been the principles of indirect utility and " correlated acquisition," dependent on ultimate physical laws. The human hand is still useful in swimming, and the fishes' fins could even be used for holding or clasping, if there were occasion for it. We might well attribute the paucity of indifferent types of structure to the agency of the rarest accidents of nature, though not in a theological sense. Animals and plants are no longer dependent for improvement on their occurrence, and, perhaps, never were after their competition and struggle for existence had fully begun. It is so much easier for them to turn to better account powers that they already possess in small degrees. Previously to such a competition and struggle, when the whole field of the inorganic conditions of life was open to simple organisms, they were doubtless much more variable than afterwards. But variability would then have been, as it is now, in no absolute sense accidental. On the contrary, variation would have been, instead of comparative stability in species, the most prominent normal feature of life. The tentative powers of life, instead of its hereditary features, trying all things, but not holding fast to that which is good, or not so firmly as afterwards, would have been its most characteristic manifestation. Our author's general difficulty in this chapter is as to how variations too small to have been of use could have been preserved, and he is correct in thinking that it could not be by Natural Selection, or the survival of the fittest, but wrong in thinking that variations are generally so rare or so insignificant, even in present forms of life as to require a power other than those of life in general to bring them forth when needed, or to produce them in useful amounts.

The first example of the working of Natural Selection is the well-known case of the neck of the giraffe. This it has been imagined, though not by Mr. Darwin, was produced by its supposed use in aiding this animal to feed on the foliage of trees, and by the occasional advantage it would give to the highest-reaching individuals, when in drought and scarcity the ground vegetation and lower foliage was consumed, and by thus enabling them to survive the others and continue the species, trans-

mitting this advantage to their offspring. Without denying
that this is an excellent hypothetical illustration of the process
of Natural Selection, Mr. Mivart attacks its probability as a
matter of fact. In reply to it he says : " But against this it
may be said, in the first place, that the argument proves too
much ; for, on this supposition, many species must have tended
to undergo a similar modification, and we ought to have at
least several forms similar to the giraffe developed from different *Ungulata*," or hoofed beasts. We would even go further
than Mr. Mivart, and hold that, on the hypothesis in question,
not only several forms, but the whole order of *Ungulata*, or
large portions of it, should have been similarly modified ; at
least those inhabiting regions subject to droughts and presenting
the alternative of grazing on the ground and browsing on the
foliage of high trees. But as these alternatives do not universally exist in regions inhabited by such animals, very long necks
would not, perhaps, characterize the whole order, if this hypothesis were true ; as the habit of herding does, for example.
We may observe, however, that this illustration from the
giraffe's neck is not an *argument* at all, and proves nothing,
though the hypothesis employed by it is very well called in
question by Mr. Mivart's criticism. But can Mr. Mivart suppose that, having fairly called in question the importance of
the high-feeding use of the giraffe's neck, he has thereby
destroyed the utility of the neck altogether, not only to the
theory of Natural Selection, but also to the animal itself ? Is
there, then, no important use in the giraffe's neck ? Is it really
the monstrosity it appears to be, when seen out of relation
to the normal conditions of the animal's life ? But if there be
any utility left in the neck, as a teleologist or a believer in
Final Causes would assume without question, and in spite of
this criticism, then it might serve the purposes of Natural
Selection even better, perhaps, than that of the mistaken
hypothesis. If our author had approached this subject in the
proper spirit, his criticism would probably have led him to an
important observation, which his desire to discredit a much
more important discovery has hidden from his view. He would
have inquired what are the conditions of existence of the
Ungulates generally, and of the giraffe in particular, which are

so close pressing, and so emphatically attest the grounds of their severest struggle for life, as to be likely to cause in them the highest degree of speciality and adaptation. The question of food is obviously not concerned in such a struggle, for this order of animals live generally upon food which is the most abundant and most easily obtained. Mr. Mivart compares his objection to one that has been made against Mr. Wallace's views as to the uses of colour in animals, "that colour being dangerous should not exist in nature," or that "a dull colour being needful, all animals should be so coloured." He quotes Mr. Wallace's reply, but does not take the clew to the solution of his difficulty respecting the giraffe's neck, which it almost forces on him. This reply was, that many animals can afford brilliant colours, and their various direct uses or values, when the animals are otherwise provided with sufficient protection, and that brilliant colours are even sometimes indirectly protective. The quills of the porcupine, the shells of tortoises and mussels, the very hard coats of certain beetles, the stings of certain other insects, the nauseous taste of brilliantly coloured caterpillars, and other instances, are given as examples. Now, what bearing has this on the long neck of the giraffe? According to our author, who is himself at this point on the defensive, it is as follows. He says: "But because many different kinds of animals can elude the observation or defy the attack of enemies in a great variety of ways, it by no means follows that there are any similar number and variety of ways for attaining vegetable food in a country where all such food other than the lofty branches of trees has been destroyed. In such a country we have a number of vegetable-feeding Ungulates, all of which present minute variations as to the length of the neck." Mr. Mivart is apparently not aware that he is here arguing, not against the theory of Natural Selection, but against a subordinate and false hypothesis under it. But if he thinks thus to undermine the theory, it must be because he is not aware of, or has not present to his imagination, the numberless ingenuities of nature, and the resources of support the theory has to rest upon. There can be no doubt that the neck of the giraffe, whatever other uses it can be put to, and it is put to several, is pre-eminently useful as a *watch-tower*. Its eyes, large and lustrous, "which beam with a peculiar mild but fearless expression, are so placed as to take in a wider range of the horizon than is subject to the vision of any other quadruped. While browsing on its favourite acacia, the giraffe, by means of its laterally projecting orbits, can direct its sight so as to anticipate a threatened attack in the rear from the stealthy lion or any other foe of the desert." When attacked, the giraffe can defend itself by powerful blows from its well-armed hoofs, and even its short horns can inflict fatal blows by the sidelong swing of its neck. But these are not its only protections against danger. Its nostrils can be voluntarily closed, like the camel's, against the sandy suffocating clouds of the desert. "The tail of the giraffe looks like an artificially constructed fly-flapper; and it seems at first incredible," says Mr. Darwin, "that this could have been adapted for its present purpose by successive slight modifications, each better and better fitted, for so trifling an object as to drive away flies; yet we should pause before being too positive, even in this case, for we know that the distribution and existence of cattle and other animals in South America absolutely depend on their power of resisting the attacks of insects; so that individuals which could, by any means, defend themselves from these small enemies, would be able to range into new pastures, and thus gain a great advantage. It is not that the larger quadrupeds are actually destroyed (except in rare cases) by flies, but they are incessantly harassed and their strength reduced, so that they are more subject to disease, or not so well enabled in a coming dearth to search for food, or to escape from beasts of prey."

This passage recalls our main problem, which does not concern the giraffe alone, but all the Ungulates; and its solution will show that this order of animals exhibits, almost as well as Mr. Wallace's examples, the resources that nature has for the protection of animals that have the disadvantage, not, indeed, generally of brilliant colours, but of exposure by living exclusively on bulky and comparatively innutritious food. Nearly all the resources of defensive warfare are exhausted in their specialties of protection. The giraffe alone is provided with a natural watch-tower, but the others are not left without defence. All, or nearly all, live in armies or herds, and some post sentinels around their herds. The numerous species of the antelope resort to natural fortifications or fastnesses. "They are the natives for the most part of the wildest and least accessible places in the warmer latitudes of the globe, frequenting the cliffs and ledges of mountain-rocks or the verdure-clad banks of tropical streams, or the oases of the desert." Other tribes depend on their fleetness, and on hiding in woods like the deer. Others, again, on great powers of endurance in flight and long marches, like the camel, with their commissaries of provision. Others, again, with powerful frames, like the rhinoceros and the bisons, resort to defensive attack. The ruminant habits and organs of large numbers are adapted to rapid and dangerous foraging, and to digestion under protection from beasts of prey and insects.

But our author, with little fertility of defence for the theory of Natural Selection, is still not without some ingenuity in attack. He objects, in the second place, that the longest-necked giraffes, being by so much the larger animals, would not be strong in proportion, but would need more food to sustain them, a disadvantage which would, perhaps, more than outbalance the long neck in times of drought; and he cites Mr. Spencer's ingenious speculations on the relations of size, food, and strength, in confirmation of this objection. But he forgets or overlooks the important physiological law of the compensation or economy of growth which prevails in variations. A longer neck does not necessarily entail a greater bulk or weight on the animal as a whole. The neck may have grown at the expense of the hind parts in the ancestors of the giraffe. If we met with an individual man with a longer neck than usual, we should not expect to find him heavier, or relatively weaker, or requiring more food on that account. But let us pass to the next illustration of the insufficiency of Natural Selection. This is the difficulty our author finds in attributing to this cause various cases of mimicry or protective resemblances of animals to other animals, or to other natural objects. In some insects this is carried to a wonderful extent. Thus, some which imitate leaves when at rest, in the sizes, shapes, colours, and markings of their wings, "extend the imitation even to the very

injuries on those leaves made by the attacks of insects or fungi."
Thus Mr. Wallace says of the walking-stick insects: "One of
these creatures, obtained by myself in Borneo, was covered
over with foliaceous excrescences of a clear olive-green colour,
so as exactly to resemble a stick grown over by creeping moss
or jungermannia. The Dyak, who brought it me, assured me
it was grown over with moss, although alive, and it was only
after a most minute examination that I could convince myself
that it was not so." And in speaking of the leaf-butterfly, he
says: "We come to a still more extraordinary part of the imitation, for we find representations of leaves in every stage of
decay, variously blotched and mildewed, and pierced with holes,
and in many cases irregularly covered with powdery black
dots, gathered into patches and spots, so closely resembling
the various kinds of minute fungi that grow on dead leaves,
that it is impossible to avoid thinking, at first sight, that the
butterflies themselves have been attacked by real fungi." Upon
these passages our author remarks: "Here imitation has
attained a development which seems utterly beyond the power of
the mere 'survival of the fittest' to produce. How this double
mimicry can importantly aid in the struggle for life seems
puzzling indeed, but much more so how the first beginnings
of the imitation of such injuries in the leaf can be developed
in the animal into such a complete representation of them; à
fortiori, how simultaneous and similar first beginnings of imitations of such injuries could ever have been developed in
several individuals, out of utterly indifferent and indeterminate
infinitesimal variations in all conceivable directions."

What ought to have been first suggested to a naturalist by
this wonderful mimicry is, what clever entomologists some insectivorous birds must have become to be able to press the
conditions of existence and the struggle for life in these insects
to such a degree of specialty. But this, after all, is not so very
wonderful, when we consider what microscopic sight these birds
must have acquired and what practice and exclusive interest
in the pursuit! We may feel pretty confident, however, that
neither Natural Selection nor any occult or transcendental
cause has ever carried protective mimicry beyond eyesight,
though it may well be a better eyesight than that even of a

skilful naturalist. There is no necessity to suppose, with our author, that the variations on which this selection depended were either simultaneous, or infinitesimal, or indifferent, for " individual differences " are always considerable and generally greatest in directions in which variations have already most recently occurred, as in characters in which closely allied races differ most from each other ; but, doubtless, a very long time was required for these very remarkable cases of mimicry to have come to pass. Their difficulties resemble those of the development of sight itself, on which our author comments elsewhere; but in these particular cases the conditions of " hide and seek " in the sport of nature present correlated difficulties, which, like acid and alkali, serve to neutralize each other. In these cases, four distinct forms of life of widely diverse origin, or very remotely connected near the beginnings of life itself, like four main branches of a tree, have come together into closest relations, as parts of the foliage of the four main branches might do. These are certain insectivorous birds, certain higher vegetable forms, the imitated sticks or leaves, certain vegetable parasites on them, and the mimicking insects. But the main phenomenon was and is the neck-and-neck race of variation and selection between the powers of hiding in the insect and the powers of finding in the bird. Our author overlooks the fact that variations in the bird are quite as essential to the process as those of the insect, and has chosen to consider elsewhere the difficulties which the developments of the eye present and in equal independence of its obvious uses. The fact that these, as well as other extraordinary cases of mimicry, are found only in tropical climates, or climates equable not only in respect to short periodic but also secular changes, accords well with the probable length of time in which this competition has been kept up; and the extraordinary, that is, rare character of the phenomenon agrees well with the probable supposition that it has always begun in what we call in science " an accident." If its beginnings were common, their natural consequences would also be common, and would not be wonderful; and if it arose from a destructive, unintelligent evil principle,—from Ahriman,—it has, at least, shown how the course of nature has been able to avoid destruction to the astonishment of human intelligence, and how Oromasdés has been able to defeat his antagonist by turning evil into good.

Let us take next our author's treatment of a supposed origin of the mammary, or milk glands:—

"Is it conceivable," he asks (p. 60), "that the young of any animal was ever saved from destruction by accidentally sucking a drop of scarcely nutritious fluid from an accidentally hypertrophied cutaneous gland of its mother? And even if one was so, what chance was there of the perpetuation of such a variation? On the hypothesis of 'Natural Selection' itself we must assume that, up to that time, the race had been well adapted to the surrounding conditions; the temporary and accidental trial and change of conditions, which caused the so-sucking young one to be the 'fittest to survive' under the supposed circumstances, would soon cease to act, and then the progeny of the mother, with the accidentally hypertrophied sebaceous glands, would have no tendency to survive the far outnumbering descendants of the normal ancestral form."

Here, as before, our author stakes the fate of the theory on the correctness of his own conceptions of the conditions of its action. He forgets, first of all, that the use of a milk gland in its least specialized form requires at least a sucking mouth, and that sucking mouths and probosces have very extensive uses in the animal kingdom. They are good for drinking water and nectar, and are used for drawing blood as well as milk; and, without reference to alimentation, are still serviceable, for *support* to parasitical animals. Might not the young, which before birth are, in a high degree, parasitical in all animals, find it highly advantageous to continue the habit after birth, even without reference to food, but for the generally quite as important use of protection against enemies, by clinging by a sucking mouth to the body of its dam? If this should cause sebaceous glands to become hypertrophied and ultimately a valuable or even an exclusive source of nutrition, it would, perhaps, be proper to describe the phenomenon as an unintended or accidental, but not as a rare or improbable one. Moreover, though on the theory of Natural Selection (or, indeed, on any theory of the continuance of a race by modifications of structures and habits), the race must, while it lives, be fitted to live, yet it need be no more fitted to do so than to survive in its offspring. No race is so well fitted to its general conditions of existence, but that some individuals are better fitted than others, and have, on the average, an advantage. And new resources do not imply abandonment of the old, but only additions to them, giving superiorities that are almost never superfluous. How, indeed, but by accidents of the rarest occurrence, could variation (much less selection) give superfluous advantages, on the whole, or except temporarily and so far as normal variations anticipate in general, regular, or usual changes in the conditions of existence? We have, to be sure, on the hypothesis, we have proposed, still to account for the original of the sucking mouth, though its numerous uses are obvious enough, on the really uniform and unvarying types of natural law, the laws of inorganic physics, the principles of suction. But we are not ambitious to rival nature in ingenuity, only to contrast its resources with those of our naturalist. His next example is a criticism of the theory of Sexual Selection. Speaking of apes, he says: "When we consider what is known of the emotional nature of these animals and the periodicity of its intensification, it is hardly credible that a female would often risk life or limb through her admiration of a trifling shade of colour or an infinitesimally greater, though irresistibly fascinating degree of wartiness." Is it credible that Mr. Mivart can suppose that the higher or spiritual emotions, like affection, taste, conscience, ever act *directly* to modify or compete with the more energetic lower impulses, and not rather by forestalling and indirectly regulating them, as by avoiding temptation in the case of conscience; or by establishing social arrangements, companionships, friendships, and more or less permanent marriages in the case of sexual preferences? All such arrangements, all grounds for the action of taste or admiration, or any but the most monstrous friendships, are prevented or removed in the lives of caged beasts. His example and his inference from it are as much as if an explorer should discover a half-famished tribe of savages sustaining life upon bitter and nauseous food, and should conclude that not only these but all savages, the most provident, or even all men, are without any choice in food, and that in providing for future wants they are influenced by no other considerations than the grossest cravings of appetite.

D

But to return to Natural Selection. The next example is that of the rattling and expanding powers of poisonous snakes. The author says that " in poisonous serpents, also, we have structures which, at all events, at first sight, seem positively hurtful to these reptiles. Such are the rattle of the rattlesnake and the expanding neck of the cobra, the former serving to warn the ear of the intended victim as the latter warns the eye." This "first sight" is all the use our author discovers in these organs; but why should these warnings be meant or used to drive away intended *victims* rather than *enemies?* Or is it among the intentions of nature to defeat those in the serpent? If the effects of such "warnings" really were to deprive these snakes of their proper food, would not experience itself and intelligence be sufficient in the wily serpent to correct such perverse instincts? It is, indeed, at first sight, curious that certain snakes, though these are the sluggish kinds, and cannot so easily escape their enemies by flight as others can, should be provided, not only with poisonous fangs, but with these means of warning either victims or dangerous enemies. But Mr. Wallace has furnished a clew to their correlation by his example of the relations between conspicuous colours and nauseous tastes in many caterpillars, the colour serving as a sign of the taste and warning birds not to touch these kinds. The poisonous fang and its use are expensive and risky means of defence; the warnings associated with them are cheap and safe. But if, as is very likely, these "warnings" are also used against intended victims, they can only be used either to paralyze them with terror or allure them from curiosity, or to produce in them that curious and paralyzing mixture of the two emotions, alarm and something like curiosity, which is all that is probably true of the supposed powers of fascination * in serpents. Perhaps, also, the rattle serves to inspire the sluggish

---

* This is a real condition of mind in the subject of it; a condition in which interest or emotion gives to an idea such fixity and power that it takes possession at a fatal moment of the will and acts itself out; as in the fascination of the precipice. It is not, however, to be regarded as a natural contrivance in the mental acquisitions of the victims for the benefit of the serpent any more than the serpent's warnings are for their benefit; but as a consequence of ultimate mental laws in general, of which the serpent's faculties and habits take advantage.

snake itself with courage ; and in this case the rattle will serve all the purposes that drums, trumpets, and gongs do in human warfare. The swaying body and vibrating tongue of most snakes, and the expanding neck and the hood of the cobras, may serve for banners. But the rattle has also been supposed to serve as a sexual call, very much as the inspirations of warfare are turned into the allurements of the tournament or as gongs also serve to call travellers to dinner. What poverty of resources in regard to the relations of *use* in the lives of animals thus distinguishes our naturalist from the natural order of things ! What wealth and capital are left for the employments and industries of Natural Selection !

In the next chapter our author charges the theory of Natural Selection with inability to account for independent similarities of structure; " that it does not harmonize with the coexistence of closely similar structures of diverse origin," like the dental structures in the dog and in the carnivorous marsupial, the Thylacine, closely similar structures and of exactly the same utilities, though belonging to races so diverse that their common ancestors could not have been like them in respect to this resemblance. But these structures really differ in points not essential to their utilities; in characters which, though inconspicuous, are marks of the two great divisions of mammalia, to which these animals belong. Our author here attacks the theory in its very citadel, and has incautiously left a hostile force in his rear. He has claimed in the preceding chapter for Natural Selection that it ought to have produced several independent races of long-necked Ungulates, as well as the giraffe ; so that, instead of pursuing his illustrations any further, we may properly demand his surrender. Of course Natural Selection requires for similar products similar means and conditions ; but these are of such a general sort that they belong to wide ranges of life; and as it does not act by "blind chance," or theological accidents, but by the invariable laws of nature and the tentative powers of life, it is not surprising that it often repeats its patterns independently of descent, or of the copying powers of inheritance.

That the highest products of nature are not the results of the mere forces of inheritance, and do not come from the birth

of latent powers and structures, seems to be the lesson of the
obscure discourse in which Jesus endeavoured to instruct Nico-
demus the Pharisee. How is it that a man can be born again,
acquire powers and characters that are not developments of
what is already innate in him? How is it possible when he
is old to acquire new innate principles, or to enter a second
time into his mother's womb and be born? The reply does
not suggest our author's hypothesis of a life turning over upon
a new "facet," or a new set of latent inherited powers. Only
the symbols, water and the Spirit, which Christians have ever
since worshipped, are given in reply; but the remarkable illustration of the accidentality of nature is added, which has been
almost equally though independently admired. " Marvel not
that I said unto thee, Ye must be born again. The wind
bloweth where it listeth, and thou hearest the sound thereof,
but canst not tell whence it cometh and whither it goeth; so is
every one that is born of the Spirit." The highest products of
nature are the outcome of its total and apparently accidental
orders; or are born of water and the Spirit, which symbolize
creative power. To this the Pharisee replied: " How can these
things be?" And the answer is still more significant: " Art
thou a master of Israel and knowest not these things?" We
bring natural evidences, " and ye receive not our witness. If I
have told you earthly (natural) things, and ye believe not, how
shall ye believe if I tell you heavenly (supernatural) things?"
The bearing of our subject upon the doctrine of Final Causes in
natural history has been much discussed, and is of considerable
importance to our author's theory and criticism.* But we propose, not only to distinguish between this branch of theology
and the theories of inductive science on one hand, but still
more emphatically, on the other hand, between it and the
Christian faith in divine superintendency, which is very liable
to be confounded with it. The Christian faith is that even the
fall of a sparrow is included in this agency, and that as men
are of more value than many sparrows, so much more is their
security. So far from weakening this faith by showing the connection between value and security, science and the theory of

* See note in Appendix.

Natural Selection have confirmed it. The very agencies that give values to life secure them by planting them most broadly in the immutable grounds of utility. But Natural Theology has sought by Platonic, not Christian, imaginations to discover, not the relations of security to value, but something *worthy* to be the source of the value considered as *absolute*, some particular worthy source of each valued end. This is the motive of that speculation of Final Causes which Bacon condemned as sterile and corrupting to philosophy, interfering, as it does, with the study of the facts of nature, or of what *is*, by preconceptions, necessarily imperfect as to what *ought to be ;* and by deductions from assumed *ends*, thought worthy to be the purposes of nature. The naturalists who " take care not to ascribe to God any intention," sin rather against the spirit of Platonism than that of Christianity, while obeying the precepts of experimental philosophy. Though, as our author says, in speaking of the moral sense and the impossibility, as he thinks, that the accumulations of small repugnances could give rise to the strength of its abhorrence and reprobation; though, as he says, " no stream can rise higher than its source ; " while fully admitting the truth of this, we would still ask, Where is its source ? Surely not in the little fountains that Platonic explorers go in search of, *à priori*, which would soon run dry but for the rains of heaven, the water and the vapour of the distilling atmosphere. Out of this come also the almost weightless snow-flakes, which, combined in masses of great gravity, fall in the avalanche. The results of moralizing Platonism should not be confounded with the simple Christian faith in Divine superintendence. The often-quoted belief of Professor Gray, " that variation has been led along certain beneficial lines, like a stream along definite lines of irrigation," might be interpreted to agree with either view. The lines on which variations are generally useful are lines of search, and their particular successes, dependent, it is true, on no theological or absolute accidents, may be regarded as being lines of beneficial variations, seeing that they have resulted through laws of nature and principles of utility in higher living forms, or even in continuing definite forms of life on the earth. But thousands of movements of variation, or efforts of search, have not succeeded to one that

has. These are not continued along evil lines, since thousands
of forms have perished in consequence of them for every one
that has survived.

The growth of a tree is a good illustration of this process,
and more closely resembles the action of selection in nature
generally than might at first sight appear; for its branches are
selected growths, a few out of many thousands that have begun
in buds; and this rigorous selection has been effected by the
accidents that have determined superior relations in surviving
growths to their supplies of nutriment in the trunk and in
exposure to light and air. This exposure (as great as is consistent with secure connection with the sources of sap) seems
actually to be sought, and the form of the tree to be the result
of some foresight in it. But the real seeking process is budding,
and the geometrical regularity of the production of buds in
twigs has little or nothing to do with the ultimate selected
results, the distributions of the branches, which are different
for each individual tree. Even if the determinate variations
really existed,—the " facets " of stable equilibrium in life, which
our author supposes,—and were arranged with geometrical
regularity on their spheroid of potential forms, as leaves and
buds are in the twig, they would probably have as little to do
with determining the ultimate diversities of life under the action
of the selection which our author admits as phyllotaxy has to
do with the branching of trees. But phyllotaxy, also, has its
utility. Its orders are the best for packing of the incipient
leaves in the bud, and the best for the exposure to light and
air of the developed leaves of the stem. But here its utility
ends, except so far as its arrangements also present the greatest
diversity of finite elements, within the smallest limits, for the
subsequent choice of successful growths; being *the nearest
approaches that finite regularity could make to* " indefinite variations in all conceivable directions." The general resemblance
of trees of a given kind depends on no formative principle other
than physical and physiological properties in the woody tissue,
and is related chiefly to the tenacity, flexibility, and vascularity
of this tissue, the degrees of which might almost be inferred
from the general form of the tree. It cannot be doubted, in
the case of the tree, that this tentative though regular budding

has been of service to the production of the tree's growth, and that the particular growths which have survived and become the bases of future growths were determined by a beneficial though accidental order of events under the total orders of the powers concerned in the tree's development. But if a rigorous selection had not continued in this growth, no proper branching would have resulted. The tree would have grown like a cabbage. Hence it is to selection and not to variation,—or rather to the *causes* of selection, and not to those of variation,—that species or well-marked and widely separated forms of life are due. If we could study the past and present forms of life, not only in different continents, which we may compare to different individual trees of the same kind, or better, perhaps, to different main branches from the same trunk and roots, but could also study the past and present forms of life in different planets, then diversities in the general outlines would probably be seen similar to those which distinguish different kinds of trees, as the oak, the elm, and the pine; dependent, as in these trees, on differences in the physical and physiological properties of living matters in the different planets,—supposing the planets, of course, to be capable of sustaining life, like the earth, or, at least, to have been so at some period in the history of the solar system. We might find that these general outlines of life in other planets resemble elms or oaks, and are not pyramidal in form like the pine, with a "crowning" animal like man to lead their growths. For man, for aught we know or could guess (but for the highly probable accidents of nature, which blight the topmost terminal bud and give ascendency to some lateral one), except for these accidents, man *may* have always been the crown of earthly creation, or always "man," if you choose so to name and define the creature who, though once an ascidian (when the ascidian was the highest form of life), *may* have been the *best* of the ascidians. This would, perhaps, add nothing to the present value of the race, but it might satisfy the Platonic demand that the race, though not derived from a source quite worthy of it, yet should come from the *best* in nature.

We are thus led to the final problem, at present an apparently insoluble mystery, of the origin of the first forms of life on the earth. On this Mr. Darwin uses the figurative language

of religious mystery, and speaks "of life with its several powers
being originally breathed by the Creator into a few forms or
into one." For this expression our author takes him to task,
though really it could mean no more than if the gravitative
properties of bodies were referred directly to the agency of a
First Cause, in which the philosopher professed to believe; at
the same time expressing his unwillingness to make hypotheses,
that is, transcendental hypotheses, concerning occult modes of
action. But life is, indeed, divine, and there is grandeur in the
view, as Mr. Darwin says, which derives from so simple yet
mysterious an origin, and "from the war of nature, from famine
and death, the most exalted object which we are capable of
conceiving, namely, the production of the higher animals."
Our author, however, is much more "advanced" than Mr.
Darwin on the question of the origin of life or archigenesis, and
the possibility of it as a continuous and present operation of
nature. He admits what is commonly called "spontaneous
generation," believing it, however, to be not what in theology
is understood by "spontaneous," but only a sudden production
of life by chemical synthesis out of inorganic elements. The
absence of decisive evidence on this point does not deter him,
but the fact that the doctrine can be reconciled to the strictest
orthodoxy, and accords well with our author's theory of sudden
changes in species, appears to satisfy him of its truth. The
theory of Pangenesis, on the other hand, invented by Mr.
Darwin for a different purpose, though not inconsistent with
the very slow generation of vital forces out of chemical actions,
—slow, that is, and insignificant compared to the normal actions
and productions of chemical forces,—is hardly compatible with
the sudden and conspicuous appearance of new life under the
microscope of the observer. This theory was invented like
other provisional theories,—like Newton's corpuscular theory of
light, like the undulatory theory of light (though this is no
longer provisional), and like the chemical theory of atoms,—for
the purpose of giving a material or visual basis to the phenomena
and empirical laws of life in general, by embodying in such
supposed properties the phenomena of development, the laws of
inheritance, and the various modes of reproduction, just as the
chemical theory of atoms embodies in visual and tangible pro-

perties the laws of definite and multiple proportions, and the relations of gaseous volumes in chemical unions, together with the principle of isomerism and the relations of equivalent weights to specific heats. The theory of Pangenesis presents life and vital forces in their ultimate and essential elements as perfectly continuous, and in great measure isolated from other and coarser orders of forces, like the chemical and mechanical, except so far as these are the necessary theatres of their actions. Gemmules, or vital molecules, the smallest bodies which have separable parts under the action of vital forces, and of the same order as the scope of action in these forces,—these minute bodies, though probably as much smaller than chemical molecules as these are smaller than rocks or pebbles, may yet exist in unorganized materials as well as in the germs of eggs, seeds, and spores, just as crystalline structures or chemical aggregations may be present in bodies whose form and aggregation are mainly due to mechanical forces. And, as in mechanical aggregations (like sedimentary rocks), chemical actions and aggregations slowly supervene and give in the metamorphosis of these rocks an irregular crystalline structure, so it is supposable that finer orders of forces lying at the heart of fluid matter may slowly produce imperfect and irregular vital aggregations. But *definite* vital aggregations and *definite* actions of vital forces exist, for the most part, in a world by themselves, as distinct from that of chemical forces, actions, and aggregations as these are from the mechanical ones of dynamic surface-geology, which produce and are embodied in visible and tangible masses through forces the most directly apparent and best understood ; or as distinct as these are from the internal forces of geology and the masses of continents and mountain formations with which they deal ; or as distinct again as these are from the actions of gravity and the masses in the solar system ; or, again, as these are from the unknown forces and conditions that regulate sidereal aggregations and movements. These various orders of molar and molecular sizes are limited in our powers of conception only by the needs of hypothesis in the representation of actual phenomena under visual forms and properties. Sir William Thomson has lately determined the probable sizes of chemical molecules from the phenomena of light and experiments re-

lating to the law of the "conservation of force." According to
these results, these sizes are such that if a drop of water were
to be magnified to the size of the earth, its molecules, or parts
dependent on the forces of chemical physics, would be seen to
range from the size of a pea to that of a billiard-ball. But
these is no reason to doubt that in every such molecule there
are still subordinate parts and structures; or that, even in
these parts, a still finer order of parts and structures exists, at
least to the extent of assimilated growth and *simple* division.
Mr. Darwin supposes such growths and divisions in the vital
gemmules; but our author objects (p. 230) that, " to admit
the power of spontaneous division and multiplication in such
rudimentary structures seems a complete contradiction. The
gemmules, by the hypothesis of Pangenesis, are the ultimate
organized components of the body, the absolute organic atoms
of which each body is composed; how then *can* they be
divisible? Any part of a gemmule would be an impossible
(because *less* than possible) quantity. If it is divisible into
still smaller organic wholes, as a germ-cell is, it must be made
up, as the germ-cell is, of subordinate component atoms, which
are then the *true* gemmules." But this is to suppose what is
not implied in the theory (nor properly even in the chemical
theory of atoms), that the sizes of these bodies are any more
constant or determinate than those of visible bodies of any
order. It is the order only that is determinate; but within it
there may be wide ranges of sizes. A billiard-ball may be divided
into parts as small as a pea, or peas may be aggregated into
masses as large as a billiard-ball, without going beyond the
order of forces that produce both sizes. Our author himself
says afterwards and in another connection (p. 290), " It is
possible that, in some minds, the notion may lurk that such
powers are simpler and easier to understand, because the bodies
they affect are so minute! This absurdity hardly bears stating.
We can easily conceive a being so small that a gemmule would
be to it as large as St. Paul's would be to us." This argument,
however, is intended to discredit the theory on the ground that
it does not tend to simplify matters, and that we must rest
somewhere in " what the scholastics called ' substantial forms.' "
But this criticism, to be just, ought to insist, not only that vital

phenomena are due to "a special nature, a peculiar innate power and activity," but that chemical atoms only complicate the mysteries of science unnecessarily; that corpuscles and undulations only hide difficulties; and that we ought to explain very simply that crystalline bodies are produced by "polarity," and that the phenomena of light and vision are the effects of "luminosity." This kind of simplicity is not, however, the purpose which modern science has in view; and, consequently, our real knowledges, as well as our hypotheses, are much more complicated than were those of the schoolmen. It is not impossible that vital phenomena themselves include orders of forces as distinct as the lowest vital are from chemical phenomena. May not the contrast of merely vital or vegetative phenomena with those of *sensibility* be of such orders? But, in arriving at *sensibility*, we have reached the very elements out of which the conceptions of size and movement are constructed,—the elements of the tactual and visual constructions that are employed by such hypotheses. Can sensibility and the movements governed by it be derived directly by chemical synthesis from the forces of inorganic elements? It is probable, both from analogy and direct observations, that they cannot (though some of the believers in "spontaneous generation" think otherwise); or that they cannot, except by that great alchemic experiment which, employing all the influences of nature and all the ages of the world, has actually brought forth most if not all of the definite forms of life in the last and greatest work of creative power.

[APPENDIX.

APPENDIX.

THE signification of the terms, use, contrivance, purpose, and intention, which are frequently somewhat loosely expressed in biological enquiries and speculations, ought to be definite and precise. So completely has the doctrine of Final Causes usurped these terms that the ambiguities of the common-sense and common-language significance of them has been overlooked by the disciples of the doctrine. Thus, Mr. Mivart in his Introduction says, "His [Mr. Darwin's] frequent use metaphorically of the expressions 'contrivance,' for example, and 'purpose,' has elicited from the Duke of Argyll and others criticisms which fail to tell against their opponent, because such expressions are in Mr. Darwin's writings merely figurative—metaphors and nothing more." But the "looseness" of the significations of the words "contrivance" and "purpose" belong to common usage irrespective of the univocal meanings that the doctrine of Final Causes seeks to impose upon them. They are like many other terms in common, or even philosophical language—like the names of colours, tastes, smells, &c.—ambiguous from being the names both of objective and subjective natures or states of facts. The word "use" is the least of all ambiguous, and refers to purely objective relations. Properly also the word "intention" is equally subjective, or denotes a state of mind in a particular mental agent, without essential reference to its objective effect. But the words "contrivance" and "purpose" are in common language ambiguous, and may refer either to an actual combination of the causes of a given effect, or to an actual imagination of the means of a given end. These words are not therefore used by Mr. Darwin figuratively, or as "metaphors and nothing more." The relations of a machine to its uses may be considered in good sound English as contrivances and purposes without thinking of what the inventor *intended*, and by omitting altogether the approximate and subjective, or imagined contrivances by which he has discovered at last the truest principles of effective and objective utility. The proper meaning of the word "intention," which really as well as etymologically signifies *concentration* and the *not* intending of something else, relates essentially to the constitution of a finite being, with limited powers. The word " intention " has no natural meaning, except relatively to those limits of power which make concentration and action on limited lines of causation advantageous. Consequently Mr. Darwin does not speak irrationally in asking, in the comparison he makes of the natural selection of variations with the selection by a builder of fragments of stone at the base of a precipice ('Animals and Plants under Domestication,' Vol. II. p. 431)—he is not irrational, we say, in asking whether " it can be reasonably maintained that the Creator intentionally ordered, if we use the words in any ordinary sense, that certain fragments of rock should assume certain shapes, so that the builder might erect his edifice ? " although as he adds, " an omniscient Creator must have foreseen every consequence which results from the laws imposed by Him." But upon this passage our author comments as follows: " Why surely every theist must maintain that in the first foundation of the universe—the primary and absolute creation—God saw and knew every purpose which every atom and particle of matter should ever subserve in all suns and systems, and throughout all coming æons of time. It is almost incredible, but nevertheless it seems necessary to think that the difficulty thus proposed rests on a sort of notion that amid the boundless profusion of Nature there is too much for God to superintend ; that the number of objects is too great for an infinite and *omnipresent* being to attend singly to each and all in their due proportions and needs." But there is indeed a difficulty, though only a lexical one, or in the consistency of meanings in words, in thinking that "attention" and "intention," meaning *concentration* upon special objects and purposes should be powers of Omniscience and Omnipotence. Though it is within the limits which theological inquiry allows itself, to ask whether Omnipotence cannot limit itself and act under the conditions specially advantageous to finite powers ; still the meanings of words ought to set limits to such inquisitiveness.

It is a contradiction of the meanings of the words " Omnipre-
sence" and " attention" to suppose an actual Omnipresent
power that should attend singly to each and all in their due
proportions and needs. But there is no objection to suppose
powers as effective as volition in each and all the movements
of nature, though these are different from our ways, and are not,
like our volitions, powers of concentration. *Superintendence*,
on the other hand, does not imply any specialty of providence,
and is what is taught by unplatonized Christianity concerning
the nature of Divinity. This teaching is in effect the same as the
primary assumption or position of experimental philosophy;
the doctrine of the universality of causation and the presumption of essential unity in nature. Even in human life the most
worshipful powers are not those of " attention" and " intention,"
however valuable these may be for the acquisition of the more
estimable, or of the superintendent habits and virtues that effect
the best without specially intending every item of detail. It is
rather with these the resulting virtues, than with those the
means of acquiring them that we most naturally associate
the divinest in nature. Surely efforts of no kind can be affirmed
of Omnipotence.

Printed in the United States
By Bookmasters